Design of Mechanical Power Transmissions

Design of Mechanical Power Transmissions

A monograph covering:
- Definitions
  force, torque, work, power
- Gear kinematics
  involute properties
  simple & compound trains
- Planetary gear trains
  kinematic analysis
    generic gear ratios
- Gear train applications
    compound hoisting block
  hybrid reduction system
- Fixed ratio transmissions
    restraint requirements
  power loss effects
- Variable ratio transmissions
  fluid coupling
  torque converter
- Case Study
  matching motor to mixer

Mechanical Design Engineering
Monograph - I

Design of Mechanical Power Transmissions

Copyright 2016
All Rights Reserved

Design of Mechanical Power Transmissions

To Louise for her understanding and support during the preparation of the mechanical design monograph series project.

CFZ

Design of Mechanical Power Transmissions

Design of Mechanical Power Transmissions

Table of Contents

Chapter 0 – Special Introduction — 1

Chapter 1 – Definitions
- Force and Torque — 5
- Rotational Speed Units — 6
- Work and Power — 7
- Torque versus Power — 8

Chapter 2 – Gear Kinematics
- Involute Geometry — 11
- Rolling Contact — 12
- Simple Gear Train — 14
- Diametral Pitch — 15
- Compound Gear Train — 16
- Alternate Approach — 18
- Compound Gear Problem — 18
- Summary — 21

Chapter 3 – Planetary Gear Trains
- Generic Planetary System — 23
- General Kinematic Rotation — 24
- Gear Ratio Equation — 26
- Train Values — 28
- Classic Ratio scenarios — 29
  - Fixed ring — 30
  - Fixed sun — 30
  - Fixed carrier — 31
  - Fixed planet — 32
- Summary — 32

- Numerical Example 32
- Geometric Restriction 33
- Automatic Transmission 34
- Generalizations 34

Chapter 4 – Gear Train Application
- Mechanical Hoisting Block 37
- Load Calculation 39
- Hybrid Planetary System 40
- Numerical Calculations 46
- Simple Gear Set Comparison 48

Chapter 5 – Fixed Ratio Transmissions
- Generic Transmission 51
- Reaction Torque 52
- Some Generalizations 53
- Case 1: Rotation in Same Direction 53
- Case 2: Rotation in Opposite Direction 54
- Example Problem 55
- Effect of Efficiency 59

Chapter 6 – Variable Ratio Transmissions
- Fluid Coupling Operation 61
- Fluid Coupling Analysis 63
- Numerical Analysis 65
- Mechanical Dynamometer Problem 68
- System Torques and Speeds 70
- Transmission Ratio 71
- Coupling Efficiency 72
- Torque Relationships 72
- Torque Calculations 73
- Power Calculations 74

Design of Mechanical Power Transmissions

- Solution Check — 75
- Torque Converter Operation — 76
- Torque Converter Analysis — 78
- Numerical Example — 78

Chapter 7 – Matching a Motor to a Load
- Introduction — 83
- Problem Statement — 84
- Mixer Characteristics — 84
- Required Mixer Torque — 85
- Motor Characteristics — 85
- Motor Torque Curves — 86
- Horse Power Calculations — 87
- Numerical Power Values — 88
- Speed Reduction — 89
- Torque Multiplication — 90
- Torque Required at Motor — 90
- Sample Calculation — 91
- Torque Calculation — 92
- Motor and Mixer Torque — 93
- Evaluation — 94
- System Starting Time — 95
- Dynamics of System — 96
- Integrating for Time — 97
- Set up for Integration — 98
- Numerical Calculations — 99
- Conclusions — 100

Design of Mechanical Power Transmissions

Design of Mechanical Power Transmissions

Preface

This is the first of a planned series of monographs dealing with Mechanical Design Engineering. Each monograph will emphasize the modeling and design performance analysis of some important physical element or aspect of the design of a mechanical device or system. Mechanical power transmission is the subject of this first monograph.

The majority of mechanical power sources are rotating devices: electric motors, internal combustion engines and jet turbines. Some form of torque versus rotational speed relationship characterizes their power output capabilities. In most applications the torque versus speed requirements of the load being driven are different from those of the power source.

This gives rise to the classic problem of the need for an intermediate device between the power source and the load to convert the motor output torque and speed to the torque and speed needs of the driven load. Whether of fixed or variable speed ratio these devices are generically classified as transmissions.

This monograph addresses this problem by providing subject content dealing with the operational kinematics and design analysis principles of the more commonly available fixed and variable ratio mechanical transmissions. Included are example applications of these principles to the transmission performance of specific power source/load combinations.

Design of Mechanical Power Transmissions

In Chapter 1 the definitions of force, torque, work and power are reviewed. Included is the generic relationship between torque, speed and power.

Chapter 2 makes use of the concept of rolling contact, characteristics of involute gear tooth geometry and diametral pitch to determine the input and output speed relationships or gear ratios in simple and compound gear systems.

In Chapter 3 the kinematic analysis process introduced in Chapter 2 is used to determine a general gear ratio relationship for a "standard" planetary gear system. Four classic gear ratio output possibilities are extracted and interpreted.

Chapter 4 presents two example applications of the previous chapters: a manual mechanical hoist that employs a planetary system and a variable high ratio hybrid planetary system.

The content of Chapter 5 covers the operational performance of generic fixed gear ratio transmission systems. Included are the effects of input/output shaft rotation and power loss on system reaction torque with specific numerical examples. .

Chapter 6 covers the analysis and operational characteristics of fluid couplings and torque converters as major variable ratio transmissions. Numerical examples are used to demonstrate their strengths and weaknesses.

In Chapter 7 a case study is presented covering the selection of a gear transmission to match an electric motor

Design of Mechanical Power Transmissions

to a rotating load under start up conditions. Both the power source output and load requirements vary with speed.

A special feature of this monograph is that in every analytic development the solution process begins with the application of fundamental engineering principles to appropriate physical models. This is followed by logical mathematical developments leading to substantive answers and results. All presentations include a combination of text explanation of the solution development together with illustrations of the symbolic mathematical process.

The goal of this monograph is to provide an understanding of the basic theory and models that are appropriate to the engineering application of the relevant subject matter in a succinct manner. It is not intended to be a textbook or comprehensive reference source. Its purpose is to assist the once acquainted reader in recalling relevant content material or to provide a concise complimentary addition to those acquiring the knowledge for the first time in a structured learning environment.

The material contained in this monograph comes from course notes in a Mechanical Design Engineering course taught at North Carolina State University by the author. An audio version of the content by chapters is also available at *http://www.designengineeringreview.com.*

Carl F. Zorowski
Raleigh, NC
September 2016

Design of Mechanical Power Transmissions

Design of Mechanical Power Transmissions

Chapter 0

Special Introduction

"The most painful thing about mathematics is how faraway you are from being able to use it after you have learned it."
James Newman - NASA
Astronaut – STS-51

Books don't normally begin with a "Chapter 0". However, it seems appropriate to insert this addition in light of the introductory quotation by James Newman. This monograph contains 100 pages of subject content. Some 30% of that number is devoted to complementary mathematical presentations to support the written text. These mathematical developments are an important part of providing a more meaningful and complete understanding of the physical concepts presented along with their application. Coupling the mathematics effectively with the text to provide this richer comprehension requires a greater degree of concentration on these manipulations then most engineers are normally willing to dedicate time to or feel comfortable with. This seems to be particularly true following their formal education and subsequent participation in professional practice. Why this appears to be so is an interesting question.

Consider for a moment our use of language. We all are fluent and comfortable with our "mother tongue" for interpersonal communication and understanding.

The ease and comfort with which this takes place is virtually automatic. It is the language in which we THINK, a consequence of having been raised "in" and "with it" from early childhood. Those who can use a language other than their "mother tongue" with some degree of fluency recognize that it, too, is a direct consequence of THINKING in terms of that language as it is spoken, read and understood. It is this "verbal/mental" process that facilitates fluency and comfort.

Mathematics is the language of engineering. It provides the symbolic means by which technical ideas and logic are conveyed and understood. To feel comfortable in the application and use of mathematics requires a similar kind of "symbolic/mental" processing. Without this capability we fall into the category of the learners referred to by James Newman. The use of mathematics then becomes painful and as a consequence is often shunned. All engineers learn mathematics as part of their formal educational process. However, the subject is often presented as a set of rules and/or operational processes to achieve some specific end. There is often insufficient emphasis on how to THINK in terms of what this language means and can convey at a deeper level. The lack of this understanding and appreciation often leads to a level of discomfort, either real or anticipated, that results in the value of mathematical developments to be overlooked and dismissed.

The level of mathematics used in this monograph is primarily algebraic manipulation. There are a few instances in which time rates of change of a variable are referred to as a derivative and the addition of cumulative periods of time as integration but these are accompanied by reassuring explanations. It is recommended that the reader exert the additional effort to follow and incorporate the mathematical developments together with the text. You will be rewarded with a richer understanding of the full benefit and value of the subject content.

Design of Mechanical Power Transmissions

Design of Mechanical Power Transmissions

Chapter 1 -Definitions

Design of Mechanical Power Transmissions deals with the generic design problem of connecting a rotational power source to the input needs of a load through some intermediate devise that matches the differences in operating characteristics of the source and the load. The specific topics covered in seven chapters includes a review of the basic definitions of torque, speed, work and power, the kinematics of simple, compound and planetary gear systems as well as fluid couplings and torque converters together with the analysis of the mechanics of generic transmission systems coupling a power source to a load.

A mechanical power transmission, shown schematically in Figure 1-1, is a physical devise placed between a rotating power source and driven machinery that changes the characteristics of the available power of the source to those required by the load.

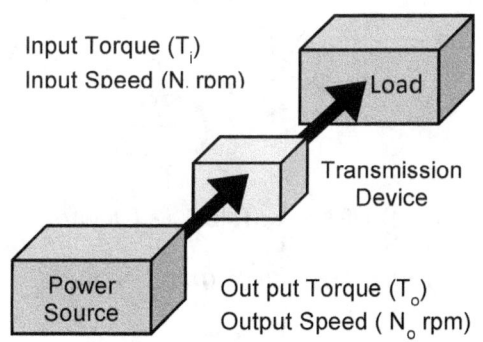

Figure 1-1   Generic Transmission System

The specific characteristics involved are torque and speed. The devise may have a speed ratio that is fixed or variable. Examples of fixed speed ratio devices are gear trains, which may be simple, compound and planetary, or hybrid. Fluid couplings and torque converters represent variable speed devices

Force and Torque

Torque is defined as the measure of the resistive twisting effect required to withstand the effect of a force attempting to rotate a body or shaft about an axis. This is depicted in Figure 1-2 in which the force F acting on the lever is being resisted by the torque T_o keeping the lever and shaft from rotating about the z-axis. From considerations of equilibrium the magnitude of T_o is simply the magnitude of the force F multiplied by the distance d measured perpendicular to the line of action of F. The units of torque are ft. lbs. or newton meters.

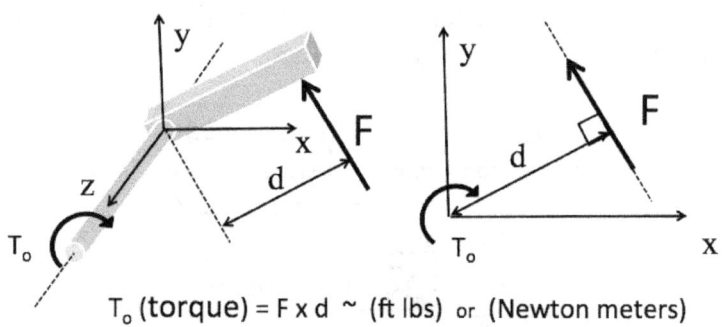

T_o (torque) = F x d ~ (ft lbs) or (Newton meters)

Figure 1-2 Force and Torque

Rotational Speed Units

Shaft rotational speed is normally expressed as N, rpm, revolutions per minute, or omega, ω, angular speed, in radians per second. It is convenient to convert from one to the other. By multiplying omega, ω, in rad/sec in Figure 1-3 by 60 min/sec and one revolution per 2π radians the conversion that one rpm is equal to 9.55 rad/sec is obtained. This is close to 10, an easy number to remember as an approximation. The inverse is one omega in rad/sec is .1047 rpm or approximately 1/10 rpm.

Conversion of customary units

$$N(rpm) = \omega\left(\frac{radians}{sec}\right)\left(\frac{60\ sec}{min}\right)\left(\frac{1\ revolution}{2\pi\ radians}\right)$$

$$N(rpm) = 9.55\ \omega\left(\frac{rad}{sec}\right)\ \ or$$

$$\omega\left(\frac{rad}{sec}\right) = .1047\ N(rpm)$$

Figure 1-3 Rotational Speed Units

Work and Power

Work performed by a rotating shaft is defined as the torque being delivered multiplied by the angle through which the shaft turns. Symbolically it is represented as T in ft. lbs. times theta, θ, in radians and has the units of ft. lbs. The rate at which work is done is defined as power. It is a measure of how fast work is being accomplished. If the toque is constant the derivative of the angular displacement theta, θ, with respect to time, t, simply becomes the angular speed omega, ω. Symbolically the power becomes the product of the T times omega, ω. A common unit of power is Horsepower equivalent to 33000 ft.lbs./min. Horsepower is related to torque in ft. lbs. and shaft speed in rpm, N, by the relationship that Horsepower is equal to 2π times N times T divided by 33,000.

Legend has it that this was the measured power output of a draft horse used to quantify the capability of the steam engine developed by Watt.

Definitions –

Work = torque x angle of shaft rotation

W = T(ft lbs) x θ (radians) = Tθ (ft lbs)

Power = rate of doing work = $\frac{d}{dt}(T \times \theta)$

if T = constant then $\frac{d}{dt}(T \times \theta) = T\frac{d\theta}{dt} = T\omega$

where ω = angular speed (radians/sec)

Horsepower : Hp = 33,000 ft lb/min

$$Hp = \frac{2\pi NT}{33000}$$

Figure 1-4 Work and Power

Torque versus Power

In most instances the torque output from a power source is not constant with speed. This gives rise to interesting relationships that can exist between torque and power. As an example consider the typical torque output of a small internal combustion engine shown in Figure 1-5. At low speed the torque is higher than at high speed and it peaks about one third through the speed range. The power output is observed to vary more in magnitude and is lower at low speed and higher at top speed with its peak near the top of its speed range. This is a consequence of power being the product of torque and speed.

Design of Mechanical Power Transmissions

Engine Speed	Torque output	Power output
rpm	ft. lbs.	Hp
500	166	15.8
1000	182	34.6
1500	187	53.4
2000	185	70.4
2500	178	84.7
3000	167	95.4
3500	152	101.2
4000	132	102.8
4500	117	100.2

Figure 1-5 Typical ICE Output

This unique relationship is illustrated by plotting the torque and the power of the internal combustion engine as a function of rotational speed. The characteristics previously described are clearly visible in Figure 1-6. The torque and the power curves do not peak at the same speed with torque peeking at a lower speed. Also the change in torque is less pronounced than the change in power.

Another way of looking at the difference between these two characteristics is that torque represents the effort to overcome a rotational load while power represents the energy required to maintain that effort continuously.

Design of Mechanical Power Transmissions

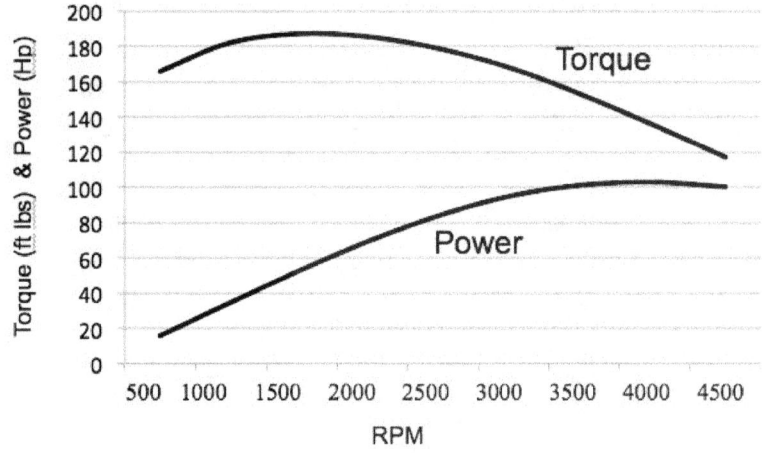

Figure 1-6   Torque / Power versus Speed

Design of Mechanical Power Transmissions

Chapter 2 – Gear Kinematics

This chapter covers the kinematics of simple and compound gear trains with an application to a generic gear selection design problem. A large variety of gear types are available and in use commercially. Some examples are spur and planetary gears commonly used for speed reduction, worm gear sets for very high reduction ratios, bevel and hypoid for changes in shaft directions and helical and herringbone for high levels of power transmission. This monograph will be limited to spur gears.

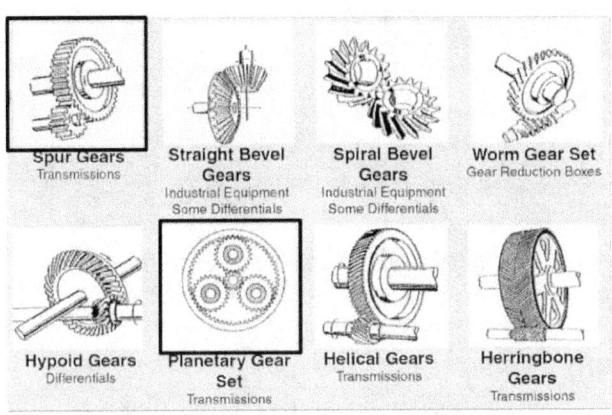

Figure 2-1: Gear Examples

Involute Geometry

The involute geometry that defines the surface curvature of a spur gear tooth possesses two very unique properties. When two gear teeth rotate over one another the point of contact remains the same distance from the centers of both gears. Also the line of action of the contact force is at a fixed angle with respect to a line joining the centers of the gears. The circle whose radius is defined by the point of contact between the gear teeth is called the pitch circle. This unique surface contact property permits the kinematics of gear trains to be analyzed as if the gears were two disks with fixed centers rolling on each other without slippage. Values of the pressure angle commonly used that define the line of action of the contact force are 14.5° and 20° degrees.

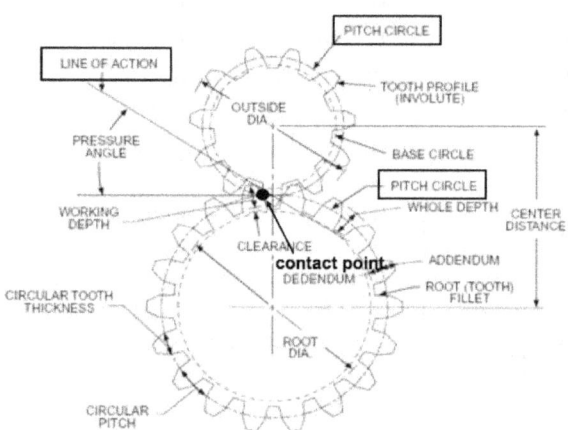

Figure 2-2: Involute Geometry

Rolling Contact

If gears are to be analyzed as disks rolling on one another it is appropriate to define what rolling contact means. As shown in Figure 2-3 the disk with center originally at point O is considered to roll with no slippage through the distance x to a new center position at O'. In the process the original radius OA turns through the angle theta, θ, to position O'A". If no slippage takes place then the distance traversed from A to A' on the flat surface must be equal to the arc length A' A".

Design of Mechanical Power Transmissions

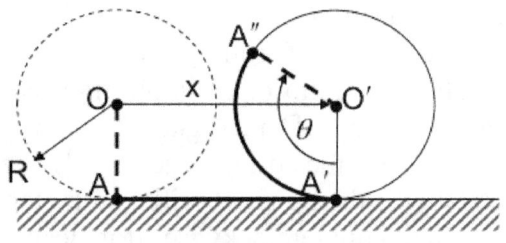

Figure 2-3: Rolling Contact

The distance moved by the center of the disk from O to O' is equal to the distance A to A' on the flat surface with both being equal to x. The arc rolled out on the disk A' to A" is simply the radius times the angle of rotation theta, θ. This results in x equal to R times θ for no slippage as shown in Figure. 2-4.

Distance moved by center of disk
　　O to O' = A to A' = x(ft)
Arc rolled out on disk
　　A' to A" = R(ft)θ(rad) = Rθ(ft)
set　O to O' = A to A' = A' to A"　so that
　　x(ft) = Rθ(ft)
differentiate with respect to time

$$\frac{dx}{dt}\left(\frac{ft}{sec}\right) = R(ft)\frac{d\theta}{dt}\left(\frac{rad}{sec}\right) \Rightarrow V\left(\frac{ft}{sec}\right) = R\omega\left(\frac{ft}{sec}\right)$$

where V = linear velocity(ft/sec), ω = andular velocity(rad/sec)

Figure 2-4 Rolling Contact Analysis

Differentiating both sides of this equation recognizing that x and θ are time variables gives the relationship V = R ω relating linear velocity, V, of the center to angular velocity, ω, of the disk.

Design of Mechanical Power Transmissions

Simple Gear Train

The concept of rolling contact will be used to determine the gear ratio of a simple train consisting of the three gears A, B and D in Figure 2-5. A is taken as the driver and D is the driven gear. If gear A rotates counterclockwise then gear B will rotate clockwise causing gear D to rotate counterclockwise. It is desired to determine the gear ratio between A and D. If gear A is rotated through an angle θ_A it will roll out an arc of R_A time θ_A on gear A. This will cause gear B to rotate by an angle of amount θ_B which is a consequence of dividing that same arc length on gear B by R_B. Thus, θ_A and θ_B are not equal since R_A is not equal to R_B.

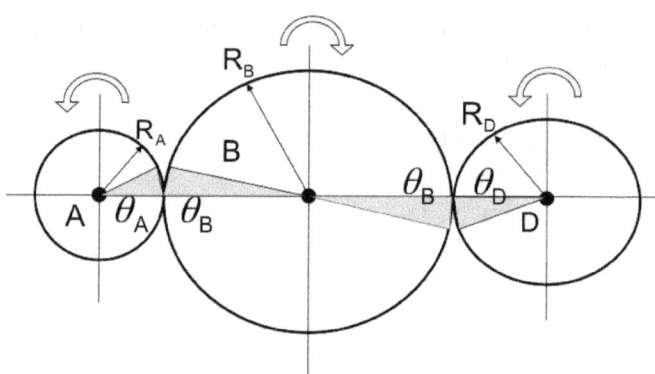

Figure 2-5: Kinematic Rotation

On the backside of gear B its rotation of θ_B rolls out an equal arc length on gear D causing gear D to rotate through an angle θ_D. Again θ_D and θ_B will not be equal since R_D and R_B are not equal. The two arc lengths on gears A and D are now set equal to each other. Equating the two arc lengths together on gears A and B gives θ_A times R_A equal to θ_B times R_B or θ_A over θ_B equal to R_B over R_A, see Figure 2-6.

As gear A rolls on gear B then

$$\theta_A R_A = \theta_B R_B \implies \frac{\theta_A}{\theta_B} = \frac{R_B}{R_A}$$

As gear B rolls on gear D then

$$\theta_B R_B = \theta_D R_D \implies \frac{\theta_B}{\theta_D} = \frac{R_D}{R_B}$$

so that $\dfrac{\theta_B}{\theta_D} \dfrac{\theta_A}{\theta_B} = \dfrac{R_D}{R_B} \dfrac{R_B}{R_A} \implies \dfrac{\theta_A}{\theta_D} = \dfrac{R_D}{R_A} = g_{AD}$ (gear ratio)

Figure 2-6 Gear Ratio Analytics

In a similar manner setting the arcs equal to each other on gears B and D gives θ_B over θ_D equal to R_D over R_B. Multiplying these two angle ratios together and eliminating θ_B and R_B gives θ_A over θ_D equal to R_D over R_A which is the gear ratio between the driving and driven gears. If the gear ratio were 3 this would indicate that gear A is three times smaller than gear D and that its speed would be three times faster than gear D. Another way of looking at this is that the gear ratio represents how much the driven gear speed is reduced from the driving gear. Note that the intermediate gear B has no effect on the gear ratio but it does change the rotational direction of gear D relative to gear A. Also since the derivative of the angular displacement, θ, is equal to the angular velocity, ω, then $d\theta/dt$ is equal to omega, ω, which in turn is equal to the speed in rpm divided by 2π. This permits the gear ratio to be expressed as the ratio of the speed of the driven gear to the speed of the driving gear.

Diametral Pitch

For the teeth on two gears to mesh together properly the circumferential arc length associated with any tooth on each gear must be the same. This circumferential arc length

Design of Mechanical Power Transmissions

is called the diametral pitch, p, and is defined as the pitch circle circumference divided by the number of teeth.

> For gear teeth to mesh they must have the same "diametral pitch" (p)
>
> $$p = \frac{2\pi R}{n}, \text{ where } n = \text{number of teeth on gear}$$
>
> so $p_A = p_D$, where $p_A = \frac{2\pi R_A}{n_A}$, $p_D = \frac{2\pi R_D}{n_D}$
>
> $\therefore \quad \frac{n_D}{n_A} = \frac{R_D}{R_A} = g_{AD}$
>
> Also since $d\theta/dt = \omega = N/2\pi$ where $N = \text{rpm}$
>
> then $\frac{\theta_A}{\theta_D} = \frac{N_A}{N_D} = \frac{R_D}{R_A} = g_{AD}$

Figure 2-7 Diametral Pitch

If the diametral pitch of two gears of different radii are set equal to one another this leads to the ratio of the number of teeth equal to the ratio of the radii which is the gear ratio, see Figure 2-7. Hence gear ratios can be determined from the number of teeth as well as the radii of the gears provided the diametral pitch is equal.

Compound Gear Train

A compound gear train, illustrated in Figure 2-8, is one in which a gear C of different radius than gear B is fixed to the shaft on which gear B rotates. In this arrangement it is gear C that drives gear D rather than gear B. It is again observed that the angular rotation of gears B and C results in the angular rotation of gear D to be in the same direction as gear A.

Design of Mechanical Power Transmissions

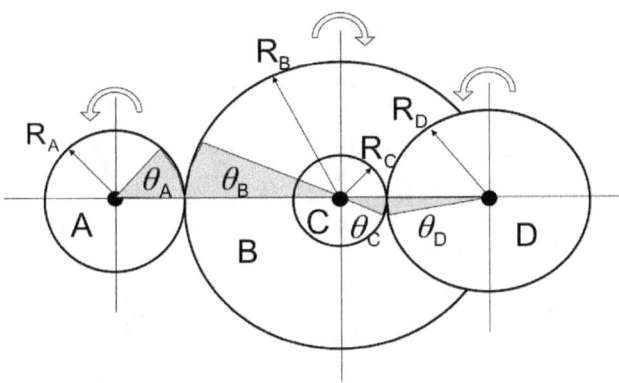

Figure 2-8: Compound Gear Train

The technique of matching arcs of contact between gears A and B and the arcs of contact between gear C and D is used to develop an equation for the gear ratio of the compound train. It is observed that the arc of contact on gear C, θ_C times R_C, is not equal to the arc θ_B times R_B on gear b even though they rotate through the same angle because their radii are different. Since gear B and gear C are on the same shaft it follows that θ_B and θ_C must be the same.

Equating the arcs on gears and A and B results in θ_B over θ_A equal to R_A over R_B as before in the simple train. Equating the arcs together on gears C and D results in θ_D over θ_C equal to R_C over R_D. Multiplying these two equations together and eliminating θ_B over θ_C which are equal to one another gives a final result of θ_A over θ_D equal to R_D over R_C times R_B over R_A, see Figure 2-9. This in turn is equal to the product of the gear ratios between gears A and B and gears C and D. In other words the gear ratios multiply in a compound train.

As gear A rolls on gear B then

$$\theta_A R_A = \theta_B R_B \Rightarrow \frac{\theta_B}{\theta_A} = \frac{R_A}{R_B}$$

As gear C rolls on gear D then

$$\theta_D R_D = \theta_C R_C \Rightarrow \frac{\theta_D}{\theta_C} = \frac{R_C}{R_D}$$

but $\theta_B = \theta_C$

so that $\dfrac{\theta_D}{\theta_C}\dfrac{\theta_B}{\theta_A} = \dfrac{R_C}{R_D}\dfrac{R_A}{R_B} \Rightarrow \dfrac{\theta_A}{\theta_D} = \dfrac{R_D}{R_C}\dfrac{R_B}{R_A} = g_{CD}g_{AB}$

Figure 2-9 Compound Gear Train Analysis

Alternate Approach-

An alternate interpretation of this result can be expressed in terms of a parameter "e" which is defined as the train value. The gear ratio g_{AD} which is the product of g_{CD} and g_{AB} is the ratio of R_D times R_B divided by R_C times R_A. This is equal to the speed of gear A over the speed of gear D. Since the diametral pitch must be the same then g_{AD} can be written as N_D over N_C times N_B over N_A.

This grouping of gear tooth numbers is defined as the reciprocal of the train value, e. Thus the train value becomes the product of the driving gear tooth numbers divided by the product of the driven gear tooth numbers as shown in Figure 2-10. This in turn is equal to the ratio of the rotational speed of the last gear over the rotational speed of the first gear. This is applicable to any simple or compound gear train or any combinations of the two.

Design of Mechanical Power Transmissions

$$g_{AD} = g_{CD}g_{AB} = \frac{R_D R_B}{R_C R_A} = \frac{N_A(\text{rpm})}{N_D(\text{rpm})}$$

since diametral pitch must be equal

$$g_{AD} = \frac{n_D n_B}{n_C n_A} = \frac{1}{e}$$ where "e" is defined as "train value"

so and the train value can be written as

$$e = \frac{\text{product of driving gear tooth numbers}}{\text{product of driven gear tooth numbers}} = \frac{N_{last}(\text{rpm})}{N_{first}(\text{rpm})}$$

where N_{last} = speed of last gear, N_{first} = speed of first gear

Figure 2-10 Definition of Train Value

Compound Gear Problem

A compound gear train is to be designed with a final gear reduction of six. Three gears having 12, 36 and 48 teeth are available. With these three select a fourth gear and the arrangement of the four gears to give the desired result. Assume any gear size is available.

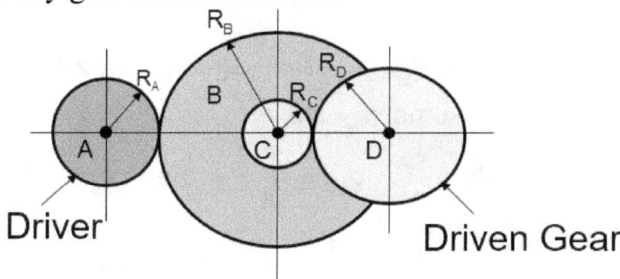

Figure 1-11: Compound Gear Problem

One solution is obtained by assuming that gears C and D have 12 and 36 teeth respectively. This means that the gear ratio between C and D is 36 divided by 12 or 3. If the total gear ratio is to be six then the gear ratio between gears A and B must be two.

Given gears with teeth – 12, 36, 48
assume $n_D = 36$, $n_C = 12$
then $\dfrac{n_D}{n_C} = g_{DC} = \dfrac{36}{12} = 3$ then $g_{AB} = 2$ and

$\dfrac{n_B}{n_A} = 2 = \dfrac{48}{n_A}$ $\Rightarrow$ $n_A = 24$ if $n_A = 48$ then $n_B = 96$

Figure 2-12 One Gear Set Solution

If gear B is chosen to have 48 teeth then gear A must have 24 teeth. However, if gear A is chosen to have 48 teeth then gear B must have 96 teeth.

An alternate solution would be to assume that C and D have 12 and 48 teeth respectively. This makes their gear ratio four. If the overall gear ratio is to be six then gears A and B must be chosen to give a gear ratio of 1.5. Selecting gear B to have 36 teeth requires gear A to have 24 teeth.

Given gears with teeth – 12, 36, 48
assume $n_D = 48$, $n_C = 12$
then $\dfrac{n_D}{n_C} = g_{DC} = \dfrac{48}{12} = 4$ then $g_{AB} = 1.5$ and

$\dfrac{n_B}{n_A} = 1.5 = \dfrac{36}{n_A}$ $\Rightarrow$ $n_A = 24$

if $n_A = 36$ then $n_B = 54$

Figure 2-13 Second Gear Set Solution

Conversely if A were chosen as the gear with 36 teeth then gear B would require 54 teeth. It is apparent that there are multiple solutions that can provide the same final gear ratio. Are there any others that would work as well?

Gear A with 24 teeth is a solution in both cases. Why?

Summary

The gear ratio of a set of spur gears A and B is simply the ratio of their radii or their number of teeth: i.e.

$$g_{ab} = R_b/R_a = N_b/N_a$$

The gear ratio of a compound spur gear set consisting of A-B and C-D with B and C on the same shaft is the product of the gear ratios of the two separate sets:

$$g_{ad} = (R_b/R_a)(R_d/R_c) = (N_b/N_a)(N_d/N_c)$$

$$g_{ad} = (g_{ab})(g_{cd})$$

… Design of Mechanical Power Transmissions

Design of Mechanical Power Transmissions

Chapter 3 – Planetary Gear Train

This chapter analyzes the behavior of a planetary gear train, develops a general equation that relates the angular rotation of its four elements, analyzes its four possible gear ratios and investigates them numerically.

Generic Planetary System

A generic planetary system consists of an internal ring gear, a planet gear, sun gear and carrier that connects the center of the sun and planet gear.

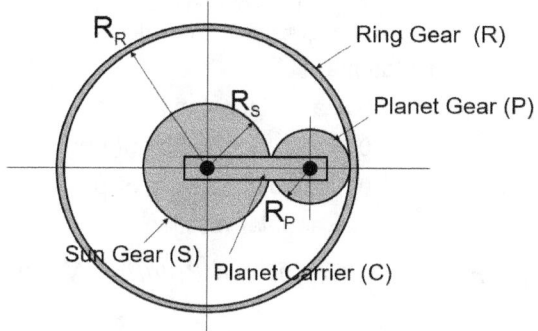

Figure 3-1: Planetary Gear Train

The diametral pitch of all the gears are the same. The centers of the sun and ring gear are the same but the sun gear can turn independently of the ring gear. The four possible gear ratios for this system include:

- Fixing the ring, rotating the sun with the carrier as the output
- Fixing the sun, rotating the carrier with the ring as the output
- Fixing the carrier, rotating the sun with the ring as the output

23

- Locking the sun to the planet and rotating the carrier with the ring as the output.

The gear dimensions are designated in terms of their radii with the subscripts s, p and r indicating sun, planet and ring. Angular rotations are expressed in terms of the angle theta using similar subscripts with the addition of theta c for the angular displacement of the carrier.

Dimensions
  R_S = radius of sun gear
  R_P = radius of planet gear
  R_R = radius of ring gear
  $R_S + R_P$ = length of carrier

Rotations
  θ_S = angular rotation of sun
  θ_R = angular rotation of ring
  θ_P = angular rotation of planet
  θ_C = angular rotation of carier

Figure 3-2: Radii and Rotation Notations

General Kinematic Rotation

The technique of equating rolled out arc lengths on matching gears for a specified input to the system and resulting output will be used to develop a relationship between all angular gear rotations. The particular inputs selected are clockwise rotation of the carrier θ_C and the ring gear θ_R as illustrated in Figure 3-3. The resulting output will be a clockwise rotation of the sun gear. To physically demonstrate the rotations of the planet and the sun gear produced by θ_C and θ_R their initial orientations are designated by their heavy dark horizontal diameters respectively in Figure 3-4.

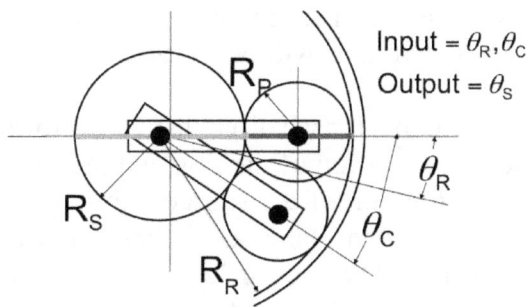

Figure 3-3: Rotational Input & Output

Consider the rotation of the planet gear first. The planet and the ring gear are initially in contact at point O. As a consequence of the ring rotation through θ_R, point O on the ring moves to point O'.

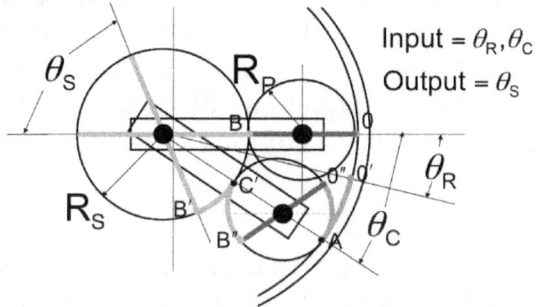

Figure 3-4: General Kinematic Rotation

With the movement of the carrier the heavy diameter on the planet rotate counterclockwise through an angle θ_P such that the point O on the planet is now at the location O".

Following rotations of the ring and carrier the planet and the ring are now in contact at point A. The arc O'A is the distance the planet has rolled out on the ring.

Design of Mechanical Power Transmissions

The arc O"A is the distance that has been rolled out on the planet. These two arcs will be set equal to each other.

Now consider what has happened to the sun gear. The clockwise rotation of the ring and the carrier produces a clockwise rotation of the sun gear through the angle θ_S. Initially the sun and planet are in contact with each other at the point B. When the carrier and ring are rotated point B on the sun gear ends up at the point B' while the point B on the planet ends up at point B".

In their final position the sun and planet gear are in contact with each other at C'. Therefore the arc rolled out on the sun gear is the circumferential distance B'C'. In a similar fashion the arc rolled out on the back side of the planet is the distance B"C'. These two arcs must be equal. Note that the arc B"C' is the same as the arc AO". It is now necessary to express all of these arc lengths in terms of gear radii and angles of rotation.

Gear Ratio Equation

The arc on the ring gear is simply the ring radius R_R times the quantity θ_C minus θ_R. The arc on the planet is the radius R_P times the quantity θ_C minus θ_P where θ_P is the counter clockwise rotation of the diameter of planet gear from its original position. The arc on the backside of the planet is equal in magnitude to the arc on the front side in contact with the sun gear. The arc on the sun gear is simply the radius R_S times the quantity θ_S minus θ_C. These arcs are all illustrated on Figure 3-5.

Design of Mechanical Power Transmissions

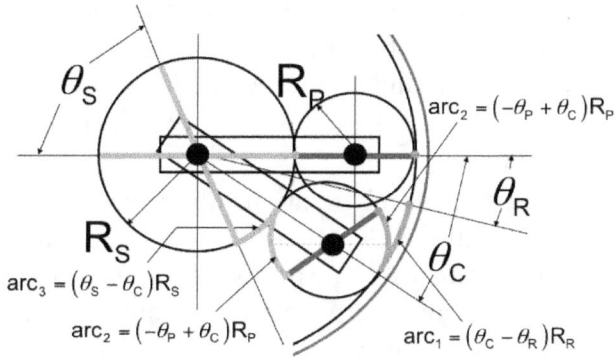

Figure 3-5: Arc Lengths

The expressions for the arcs between the planet and ring and the two arcs between the planet and sun are now set equal to each other as illustrated in Figure 3-6. Combining these two equations and eliminating the R_P time θ_C minus θ_P term results in an equation involving only the radii of the sun, ring and their angular rotations along with that of the carrier.

Equating Arcs

arc 1 = arc 2 (red)

$$(\theta_C - \theta_R)R_R = (-\theta_P + \theta_C)R_P$$

arc 2 = arc 3 (green)

$$(-\theta_P + \theta_C)R_P = (\theta_S - \theta_C)R_S$$

Combine and eliminate $(-\theta_P + \theta_C)$

$$(\theta_C - \theta_R)R_R = (\theta_S - \theta_C)R_S$$

or $\theta_S = \theta_C\left(1 + \dfrac{R_R}{R_S}\right) - \theta_R\left(\dfrac{R_R}{R_S}\right)$

Figure 3-6: Final Angular Rotation Equation

Solving this equation for θ_S as the output gives θ_C times the quantity one plus the ratio of R_R to R_S minus θ_R

times the quantity R_R over R_S. This one equation relates all possible rotations between the ring, sun and carrier. Note that the planet radius is not included.

Train Values

This general relationship between the angular rotations of all the elements can also be obtained using the concept of train values. The specific train value for e for a planetary system in terms of N_R, N_S and N_C is given by the difference between N_R and N_C divided by the difference between N_S and N_C, see Figure 3-7.

Now if the carrier is fixed but the ring and sun are permitted to rotate then the train value e is given by the negative of N_R over N_S as defined by the earlier discussion of train value.

Train value "e" for planetary system is defined as

$$e = \frac{N_R - N_C}{N_S - N_C} \text{ with } e = -\frac{N_R}{N_S}$$

(sun and ring rotate in opposite direction)

so that

$$\frac{N_R}{N_S} = \frac{N_C - N_R}{N_S - N_C} \Rightarrow N_S - N_C = \frac{N_S}{N_R}(N_C - N_R)$$

$$N_S = N_C\left(1 + \frac{N_S}{N_R}\right) - N_R\left(\frac{N_S}{N_R}\right)$$

or $\theta_S = \theta_C\left(1 + \frac{R_R}{R_S}\right) - \theta_R\left(\frac{R_R}{R_S}\right)$ same as previously

Figure 3-7: Train Values

The minus sign is introduced because the rotation of the ring and sun will be in different directions since the ring gear is an internal gear. Setting these two train values equal to one another, solving for N_S and replacing the number of teeth by angles of rotation outside the bracket terms results in the same general equation previously developed. However, it doesn't explain how the physical kinematics of the system take place. Not easy to apply to a hybrid system.

Classic Scenarios

The four classic gear ratio arrangements for a planetary system are listed in Figure 3-8 in terms of which element is fixed, which is the input and which is the output. Each of these will be analyzed in terms of the general equation relating the rotations of the sun, ring and carrier and radii of the sun and ring gears.

	Ring	Sun	Carrier	Planet
Case 1	fixed	driver	driven	
Case 2	driven	fixed	driver	
Case 3	driven	driver	fixed	
Case 4	driver	driven		fixed

Figure 3-8: Four Classic Gear Ratios

In **Case 1,** Figure 3-9, the ring is fixed. Setting θ_R equal to zero in the general solution and considering the carrier as being driven by the sun gives the ratio of θ_S over θ_C as the quantity 1 plus R_R over R_S. Since this quantity is always positive the sun and carrier turn in the same direction. With R_R over R_S always greater than one the sun will turn faster than the carrier. This means there is a gear reduction from the sun to the carrier or from the input to the output.

With ring gear fixed set $\theta_R = 0$ in

$$\theta_S = \theta_C\left(1 + \frac{R_R}{R_S}\right) - \theta_R\left(\frac{R_R}{R_S}\right) \quad \text{to give}$$

$$\theta_S = \theta_C\left(1 + \frac{R_R}{R_S}\right)$$

or $\quad \dfrac{\theta_S}{\theta_C} = \left(1 + \dfrac{R_R}{R_S}\right)$

Figure 3-9: Case 1 Gear Ratio

In **Case 2**, Figure 3-10, the sun is fixed. Setting θ_S equal to zero with the carrier acting as the driver and the ring the driven gear gives the ratio θ_C over θ_R as the quantity one over one plus R_S over R_R.

With sun gear fixed set $\theta_S = 0$ in

$$\theta_S = \theta_C\left(1 + \frac{R_R}{R_S}\right) - \theta_R\left(\frac{R_R}{R_S}\right) \quad \text{to give}$$

$$0 = \theta_C\left(1 + \frac{R_R}{R_S}\right) - \theta_R\left(\frac{R_R}{R_S}\right)$$

or $\quad \dfrac{\theta_C}{\theta_R} = \dfrac{1}{\left(1 + \dfrac{R_S}{R_R}\right)}$

Figure 3-10 Case 2 Gear Ratio

With the right side of the equation positive the ring and carrier will turn in the same direction. Since R_S over R_R will always be between one and zero the right side of the equation will always be less than one. This means the gear ratio is less than one and the ring gear will turn faster than the carrier. This might be classified as an overdrive gear ratio where the input is turning slower than the output.

Design of Mechanical Power Transmissions

In **Case 3**, Figure 3-11, the carrier is fixed. Setting θ_C equal to zero with the sun being the driver and the ring being driven gives the ratio of θ_S over θ_R equal to minus R_R over R_S.

With carrier fixed set $\theta_C = 0$ in

$$\theta_S = \theta_C \left(1 + \frac{R_R}{R_S}\right) - \theta_R \left(\frac{R_R}{R_S}\right) \quad \text{to give}$$

$$\theta_S = -\theta_R \left(\frac{R_R}{R_S}\right)$$

or $\quad \dfrac{\theta_S}{\theta_R} = -\left(\dfrac{R_R}{R_S}\right)$

Figure 3-11: Case 3 Gear Ratio

The negative sign indicates that the sun and the ring will turn in opposite directions. With R_R over R_S greater than one the sun will turn faster than the ring giving another speed reduction from the input to the output.

Case 4, Figure 3-12, is the easiest to analyze. If the planet is fixed to either the carrier or ring gear there will be no relative motion between any of the system components.

$$\theta_R = \theta_S$$

$$\left(\frac{\theta_S}{\theta_R}\right) = 1$$

Figure 3-12: Case 4 Gear Ratio

The entire unit turns together which results in θ_R equal to θ_S. Therefore the gear ratio between the input and output is one. There is no reduction. The system acts as if it were a simple rotating shaft.

Design of Mechanical Power Transmissions

Summary

Figure 3-13 lists the gear ratios for the four cases considered where the letters R and S represent either the radii of the ring and sun or the number of teeth on each. Three of the four gear ratios indicate that input and out rotations are the same while Case 3 indicates they will be reversed. It is again observed that no planet geometric parameters are involved although the system won't function without one.

	Input	Output	Stationary	Gear Ratio
Case 1	Sun-S	Carrier-C	Ring-R	1+R/S
Case 2	Carrier-C	Ring-R	Sun-S	1/(1+S/R)
Case 3	Sun-S	Ring-R	Carrier-C	-R/S
Case 4	Sun-S	Ring-R	Planet-P	1

Figure 3-13: Summary of Gear Ratios

Numerical Example

Consider a numerical example in which the ring gear has 75 teeth and the sun has 35. Calculating the four gear ratios gives the results in Figure 3-14. Case 1 provides a significant reduction in speed. Case 2 results in a slight increase in speed of the output over the input. Case 3 gives a significant reduction in speed in the opposite direction to the input and Case 4 is the same as replacing the system with a simple shaft. This combination is much like a three speed transmission in an automobile. Case 1 is first gear, Case 4 is second gear, Case 2 is third gear and Case 3 is reverse.

Design of Mechanical Power Transmissions

N_R=75 teeth N_S=35 teeth

	Input	Output	Stationary	Gear Ratio	
Case 1	Sun-S	Carrier-C	Ring-R	1+R/S	3.14
Case 2	Carrier-C	Ring-R	Sun-S	1/(1+S/R)	0.68
Case 3	Sun-S	Ring-R	Carrier-C	-R/S	-2.14
Case 4	Sun-S	Ring-R	Planet-P	1	1

Figure 3-14: Example Gear Ratios

Multiplying these numerical ratios by the reduction in the differential of the automobile would be about right.

Geometric Restriction

Although the planet size does not impact the gear ratios there is a system geometric restriction that must be satisfied. The radius of the ring gear is equal to the radius of the sun gear plus the diameter of the planet. It then follows that the number of teeth on the ring must be equal to the number of teeth on the sun plus 2 times the number of teeth on the planet as shown in Figure 3-15

From geometry of system

$$R_R = R_S + 2R_P$$

and since diametral pitch must be the same for all gears then

$$n_R = n_S + 2n_P \quad \text{or}$$

$$n_P = \frac{(n_R - n_S)}{2}$$

Thus $(n_R - n_S)$ must be an even number so that n_p is a whole number of teeth.

Figure 3-15: Restriction On Planet Size

Solving this equation for the teeth on the planet gives the difference between the number of teeth on the ring and on the sun divided by 2. This in turn means the difference between the number of teeth on the ring and the sun must be an even number so that the number of teeth on the planet is a whole number.

Automatic Transmission

Figure 3-16 is a cutaway of an automatic automotive transmission showing two planetary systems working in series with each other. By proper selection of gear sizes for each and appropriate selections of inputs and outputs across the combined system a large

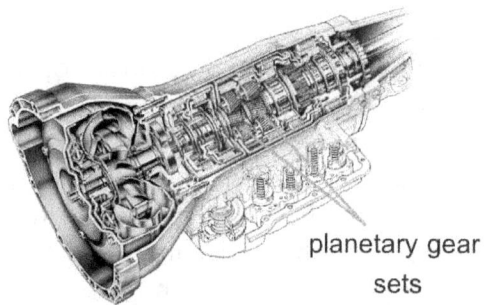

Figure 3-16: Automatic Transmission

choice of gear ratios can be achieved both forward and in reverse not all of which would necessarily be used.

Generalizations

A planetary gear system can provide three distinct gear ratios depending which component is fixed. These include a reduction in the same direction, an overdrive in the same direction and a reduction in the opposite direction.

Design of Mechanical Power Transmissions

1. Planetary gear system provides three distinct gear ratios depending on which component is fixed:
    a. reduction in same direction
    b. overdrive in same direction
    c. reduction in opposite direction
2. Gear ratios are only dependent on the size of sun and ring gears, planet size no effect.
3. Difference in number of teeth of ring and sun gear must be an even number.

Figure 3-17: System Generalizations

The gear ratios are only dependent on the size of the sun and ring gears. The planet has no effect. The difference in number of teeth on the ring and sun gear must be an even number. These generalizations are all summarized in Figure 3-17.

Design of Mechanical Power Transmissions

Design of Mechanical Power Transmissions

Chapter 4 – Planetary Gear Examples

This chapter covers the application of the planetary gear system analysis to a mechanical hoist problem and the analysis of a hybrid gear reduction composed of two planetary systems coupled together with a compound gear set.

Mechanical Hoisting Block

The cutaway of a mechanical hand operated chain hoist is shown in Figure 4-1. The hand chain sprocket wheel is 12 inches in diameter and the load chain sprocket is 6 inches in diameter. Gear a is keyed to the hand crank wheel shaft and has 15 teeth. Gears b (35 teeth) are free to rotate on a carrier that drives the load chain sprocket. The internal ring gear c is fixed to the inside of the hoist casing.

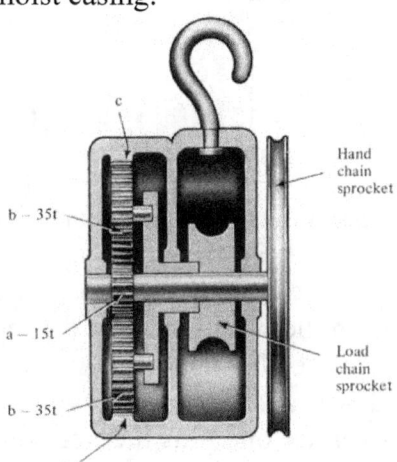

Figure 4-1: Mechanical Hoisting Block

The problem is to determine the gear ratio between the hand crank sprocket and the chain sprocket

37

and the load that can be lifted with a pull of 100 lbs. on the hand chain.

The geometry of the system is used to determine the number of teeth on the internal gear fixed to the casing. The radius of gear c must be equal to the radius of gear a plus 2 times the radius if gear b. Since the diametral pitch of all gears must be the same the number of teeth on the gears will be in the same relationship as that of the radii. With n_a given as 15 and n_b equal to 35 the number of teeth on gear c is calculated to be 85, see Figure 4-2.

From geometry of system
$$R_c = R_a + 2R_b$$
and since diametral pitch must be the same for all gears then
$$n_c = n_a + 2n_b$$
but $n_a = 15$ and $n_b = 35$
so $n_c = 15 + 2(35) = 85$ teeth

Figure 4-2: Teeth on Ring Gear

The internal casing gear a is the ring gear of the planetary and is fixed. Gear b which is connected directly to the hand crank sprocket is the driver and the sun gear in the planetary. The carrier which is connected directly to the chain sprocket is the output. With the sun as the driver and the carrier as the output the gear ratio g_{sc} is given by one plus the ratio of the number of teeth of the ring to the sun. Substituting the appropriate numbers into this relationship gives a numerical gear ratio of 6.7.

Design of Mechanical Power Transmissions

$$g_{sc} = 1 + \frac{R}{S} = 1 + \frac{n_c}{n_a} = 1 + \frac{85}{15} = 6.7$$

Figure 4-3: Total Gear Ratio

The direction of rotation of the hand chain sprocket to the load chain sprocket will be the same and the hand sprocket must be rotated 6.7 turns to produce one rotation of the load chain sprocket.

Load Calculation

Assuming there is minimal friction in the system and losses can be neglected the work done "on the system" must be equal to the work done "by the system". That is the "work in" is equal to the "work out". The work is expressed as the product of the input torque times the angular rotation of the hand crank sprocket while the work out will be the output torque times the corresponding angular rotation of the load chain sprocket.

$$T_{hs}\theta_a = T_{ls}\theta_b \Rightarrow T_{ls} = T_{hs}\frac{\theta_a}{\theta_b}$$

but $\quad \frac{\theta_a}{\theta_b} = g_{sc} = 6.7 \quad T_{hs} = F_h R_h \quad$ and $\quad T_{ls} = F_l R_l$

finally then

$$F_l = F_h \frac{R_h}{R_l} g_{sc} = 100\left(\frac{6}{3}\right)6.7 = 1340 \text{ lb force}$$

Figure 4-4: Load Calculation

It is also recognized that the input torque can be expressed as the force on the hand chain multiplied by the hand crank sprocket radius. In a similar fashion the output torque can be expressed as the load force times the radius of the load chain sprocket. Combining these relationship gives the hand chain force F_I equal to the

Design of Mechanical Power Transmissions

load chain force F_h times the ratio of the hand crank sprocket radius R_h to the radius of the load chain sprocket R_l multiplied by the gear ratio g_{sc}. A force of 100 lbs. lifts a load of 1340 lbs. as illustrated in Figure 4-4.

Hybrid Planetary Systems

A novel hybrid planetary system consisting of two planetary systems uniquely connected together is shown in Figure 4-5.

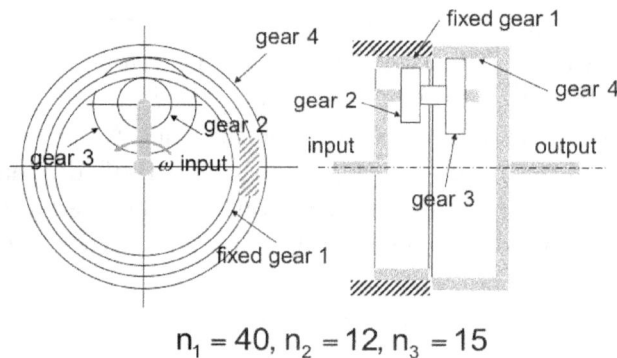

$n_1 = 40, n_2 = 12, n_3 = 15$

Figure 4-5: Hybrid Planetary System

This hybrid planetary can achieve very high gear ratios useful in making fine adjustments of rotary positioning mechanisms. In this unit gear 1 is fixed while the carrier of the left unit is the input. The left planet, gear 2, is fixed to a shaft that also drives the planet, gear 3, of the right planetary system. The ring gear 4 is the output.

For the gear sizes given the problem is to determine the gear ratio between the input and output

Design of Mechanical Power Transmissions

shaft, the direction of the output relative to the input and how the gear ratio changes as the size of gear 3 approaches the size of gear 2.

The technique of displacing the input by a finite rotation, graphically showing the arc of contact rolled out on each gear and equating equal arcs of contact is used to develop a general relationship between angular rotations and sizes of the systems elements.

Refer to Figure 4-6 and begin with a clockwise rotation of the carrier of gear 2, shown dotted through an angle θ_{in}. The planet and ring gear of the left unit is shown in a bolder black outline. The vertical on gear 2 will rotate counter clockwise. The initial point of contact O between gear 2 and ring gear 1 which is fixed ends up at O' on gear 2.

The final point of contact between gear 2 and the ring gear will be at point O". The arc rolled out on the ring gear will be from O to O" while the arc rolled out on gear 2 will be from O' to O''. These two arcs must be equal in length.

Design of Mechanical Power Transmissions

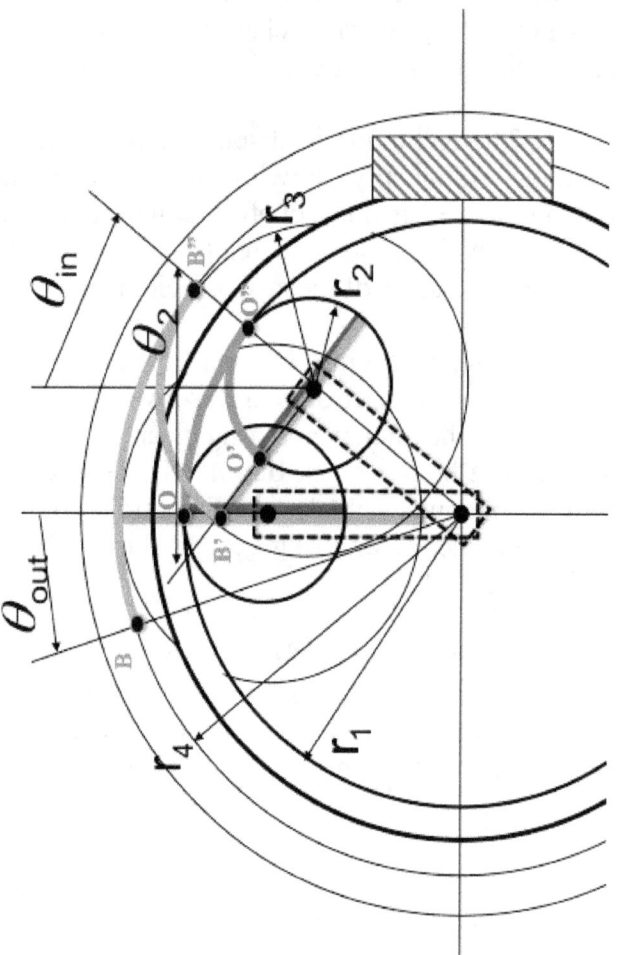

Figure 4-6 Kinematics of Rotation

Design of Mechanical Power Transmissions

With gear 2 and gear 3 on the same shaft the diameter on gear 3 will rotate through the same counterclockwise rotation as gear 2. Its initial point of contact with ring gear 4 will be vertically above the initial point of contact between gear 2 and ring gear 1. This initial point of contact will move to point B' at the end of the diameter.

It is now assumed that gear 4, the output, will rotate through the angle θ_{out} as a consequence of the input to the carrier. Its initial point of contact with gear three will now move to location B. Since the final point of contact between gear 3 and gear 4 will be at B" the arc rolled out on the output gear 4 will be from B to B". In a similar fashion the arc rolled out on gear 3 will be from B' to B". Again these arcs must be equal to one another.

The arc lengths now need to be expressed in terms of gear radii and their angular rotations. Refer to Figure 4-7 for their determination.

The arc O'O" on gear 2 is r_2 times θ_2 which is the angle between the final orientation of the carrier and the final position of the diameter on gear 2. The arc OO' on the fixed ring gear 1 is simply the product of r_1 times the input angle θ_{in}.

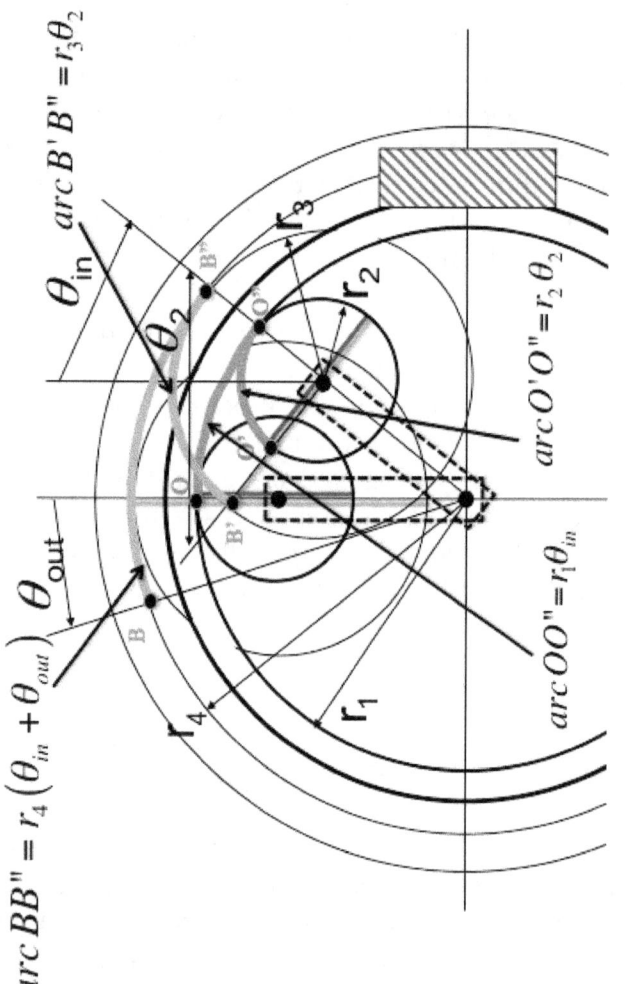

Figure 4-7 Arc Lengths

Design of Mechanical Power Transmissions

The arc BB' is given by r_3 times θ_2 while the arc BB'' on the output gear 4 is r_4 times the quantity $\theta_{in} + \theta_{out}$. θ_{in} and θ_{out} are added in this expression as θ_{out} is assumed to be positive in the counterclockwise direction.

Arcs on Gears 1 and 2
 gear 2: $r_2\theta_2$, gear 1: $r_1\theta_{in}$
set equal: $r_2\theta_2 = r_1\theta_{in}$
Arcs on Gears 3 and 4
 gear 3: $r_3\theta_2$, gear 4: $r_4(\theta_{in} + \theta_{out})$
set equal: $r_3\theta_2 = r_4(\theta_{in} + \theta_{out})$

Figure 4-8: Arcs on Gear Set

The arc length expressions are now equated in Figure 4-9. This gives $r_2\,\theta_2$ equal to $r_1\,\theta_{in}$ for the input side of the system. For the output side $r_3\,\theta_2$ is equal to r_4 times the quantity θ_{in} plus θ_{out}. These two equations can now be combined by eliminating the angle θ_2 between them.

Kinematic Equations
$$r_2\theta_2 = r_1\theta_{in}, \quad r_3\theta_2 = r_4(\theta_{in} + \theta_{out})$$
Eliminating θ_2
$$\left(\frac{r_1}{r_2}\right)\theta_{in} = \left(\frac{r_4}{r_3}\right)(\theta_{in} + \theta_{out})$$
Solve for θ_{out}:
$$\theta_{out} = \theta_{in}\left(\left(\frac{r_1}{r_2}\right)\left(\frac{r_3}{r_4}\right) - 1\right) = \theta_{in}\left(\left(\frac{n_1}{n_2}\right)\left(\frac{n_3}{n_4}\right) - 1\right)$$

Figure 4-9: Final Gear Ratio

Design of Mechanical Power Transmissions

The input to output ratio is now obtained by the elimination of θ_2 in Figure 4-9. The result is r_1 over r_2 times θ_{in} equal to r_4 over r_3 times the quantity θ_{in} plus θ_{out}. Solving for θ_{out} gives the result θ_{in} times the product of the ratio of r_1 over r_2 times r_3 over r_4 minus 1. The radii in this expression can be replaced by the number of teeth on the respective gears to give the final equation for the gear ratio.

Numerical Calculations

The numerical data given in the original problem statement can now be applied to determine the gear ratio. This first requires that the number of teeth on gear 4 be determined. It is recognized from the geometry of the system that r_1 minus r_2 must be equal to r_4 minus r_3. Substituting into this relation the number of teeth on gears 1, 2 and 3 as 40, 12 and 15 results in gear 4 possessing 43 teeth as indicated in Figure 4-10.

When the numbers of all the gear teeth are substituted into the relationship between θ_{out} and θ_{in} the resulting gear ratio of the system is 6.14. Since the output rotation was assumed opposite to the input rotation and the gear ratio came out positive the assumed direction for the output was correct. In other words the input and output shafts rotate in opposite directions.

Design of Mechanical Power Transmissions

$$r_1 - r_2 = r_4 - r_3$$
or $\quad r_4 = r_1 - r_2 + r_3 \quad \Rightarrow \quad n_4 = n_1 - n_2 + n_3$
$$n_4 = 40 - 12 + 15 = 43$$

and $\quad \theta_{out} = \theta_{in}\left(\left(\dfrac{n_1}{n_2}\right)\left(\dfrac{n_3}{n_4}\right) - 1\right) = \theta_{in}\left(\left(\dfrac{40}{12}\right)\left(\dfrac{15}{43}\right) - 1\right)$

or $\quad \theta_{out} = \theta_{in}(1.16 - 1) = 0.16\, \theta_{in}$

so that $\quad g = \dfrac{\theta_{in}}{\theta_{out}} = \dfrac{1}{0.16} = 6.14$

Figure 4-10: Numerical Gear Ratio

As a special case consider what happens if gear 2 and gear 3 are the same size. That is n_2 equals n_3. This in turn requires that n_4 be 40. When these gear tooth numbers are substituted into the gear ratio equation θ_{in} over θ_{out} is equal to infinity as indicated in Figure 4-11. Physically this corresponds to a zero output irrespective of the input speed. The output shaft remains stationary no matter what the input speed is.

If $n_3 = n_2$ then n_4 will also change
assume $n_3 = n_2 = 12$
then $\quad n_4 = n_1 - n_2 + n_3$
$$n_4 = 40 - 12 + 12 = 40$$

and $\quad \theta_{out} = \theta_{in}\left(\left(\dfrac{n_1}{n_2}\right)\left(\dfrac{n_3}{n_4}\right) - 1\right) = \theta_{in}\left(\left(\dfrac{40}{12}\right)\left(\dfrac{12}{40}\right) - 1\right)$

or $\quad \theta_{out} = \theta_{in}(1 - 1) = 0 \times \theta_{in}$

so that $g = \dfrac{\theta_{in}}{\theta_{out}}$ is infinite, i.e. there will be no output
irrespective of the input speed.

Figure 4-11: Special Case

What would be observed is that gears 2 and 3 simply run around the inside of both ring gears, which in turn remain stationary?

It is of interest to look at what happens as n_3 is reduced in number approaching the value of n_2 while keeping n_2 and n_1 the same. From the tabulated figures in Figure 4-12 it is seen that the system gear ratio increase dramatically for the intermediate figures of n_3 equal to 14 and 13.

n_1	n_2	n_3	n_4	g
40	12	15	43	6.14
40	12	14	42	9.00
40	12	13	41	17.57
40	12	12	40	infinite

Figure 4-12: Gear Ratio Comparison

Simple Gear Set Comparison

The compactness of the hybrid planetary system compared to a simple gear set is illustrated in Figure 4-13 in this side by side scale rendition in which the size of the driver gears are the same. The gear tooth numbers for the gear ratio of 17.6 from the previous example are as listed. It is observed that the diameter of the output gear on the simple gear set is more than 5 times larger than the diameter of the input ring gear n the hybrid system.

Design of Mechanical Power Transmissions

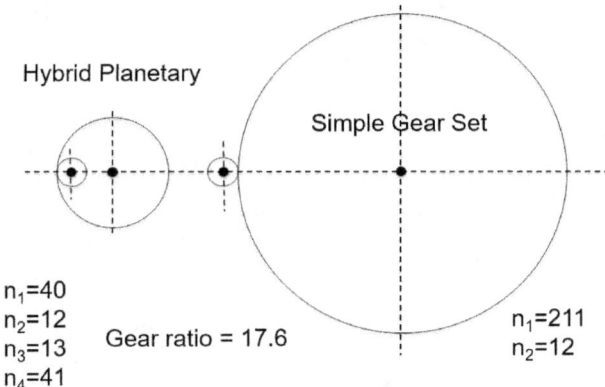

Hybrid Planetary

Simple Gear Set

$n_1=40$
$n_2=12$
$n_3=13$
$n_4=41$

Gear ratio = 17.6

$n_1=211$
$n_2=12$

Figure 4-13: Gear System Comparison

Design of Mechanical Power Transmissions

Design of Mechanical Power Transmissions

Chapter 5 – Fixed Ratio Transmissions

This chapter covers the analysis of fixed gear systems used to match power source characteristics to load requirements both with and without consideration of internal power losses.

Generic Transmission

Consider the generic fixed gear system depicted in Figure 5-1 with the input shaft rotating clockwise and an input speed of ω_i. It will be assumed that the output shaft is rotating in the same direction with an output speed of ω_o. The input shaft is delivering torque from the power source to the transmission while the output shaft is providing the torque required by the load.

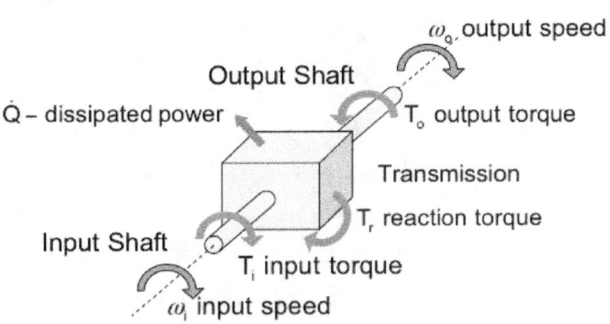

Figure 5-1: Generic Transmission

The input torque from the power source is designated T_i and is in the same direction as the rotation of the input shaft while the load torque resisting the rotation of the output shaft is designated T_o and will be directed opposite to the direction of rotation of the output shaft. With the input and output shafts rotating in the same direction T_i and T_o will be in opposite directions.

Assuming that the fixed gear transmission has a finite gear ratio then the magnitude of T_i and T_o will be different. To satisfy rotational equilibrium of the transmission casing will require a reaction torque be applied to the gearbox to hold it stationary. If energy is dissipated as heat internal to the gearbox by a lubricant to eliminate gear and bearing wear it will need to be extracted from the system.

Reaction Torque

The transmission shown in Figure 5-2 contains three gears as indicated by the input and output shafts and the intermediate bearing cover plate.

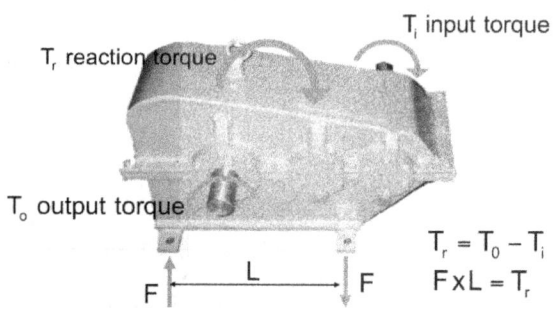

Figure 5-2: Example transmission

The input and output shafts will rotate in the same direction while the input and output torques will be in opposite directions. To hold the transmission stationary a reaction torque, T_r, must be applied to the casing. It will be equal to the difference between the output and the input torques. T_r is equal to $T_o - T_i$. This is actually applied through the fastening of the transmission feet to some external surface by the forces F whose magnitude will be T_r divided by the distance L between the forces.

Some Generalizations
- The input speed and rotation, ω_i, and input torque, T_i, on transmission are always in the same direction.
- The output speed and rotation, ω_o, and output torque, T_o, on the transmission are always in the opposite direction.
- The transmission reaction torque, T_r, is dependent on the magnitude and direction of the input and output torques.
- Power transmitted through the device is assumed to be conserved if there are no internal losses.
- The gear or speed ratio is defined as ω_i/ω_o.

The generic transmission problem is one of matching the power source torque and speed to the torque and speed requirements of the load. Some factors that can affect and complicate the analysis of this generic problem include the variable nature of power source or load requirements with speed and time to achieve operating conditions and the efficiency of the transmission. In the examples that follow it will be assumed for simplicity that power and load requirements do not vary with speed however system efficiency will be included.

Case 1: Rotation in Same Direction
Consider the case of the output shaft and the input shaft turning in the same direction with no power loss in the transmission. Equating the power in to the power out results in the output torque, T_o, being equal to the input

torque, T_i, times the gear ratio, g. In other words a torque multiplication takes place.

power in = power out

$$T_i\omega_i = T_o\omega_o \Rightarrow T_o = T_i\left(\omega_i/\omega_o\right)$$

but $\left(\omega_i/\omega_o\right) = g$ speed ratio

so $T_o = T_i g$ if $g > 1$ then T_o is multiplied

and $T_r = T_o - T_i = T_i(g-1)$

Direction of T_r is the same as the input shaft

Figure 5-3: Rotation in same direction

With the input and output torques on the transmission being in opposite directions the reaction torque on the casing will be given by the input torque minus the output torque or the input torque times the quantity the gear ratio minus one, see Figure 5-3. For gear ratios greater than one the reaction torque and input torque are in the same direction. If the gear ratio is less than one, a condition of overdrive, then the direction of the reaction torque is opposed to the direction of the input torque. If g is one then T_r is zero and the transmission really has no effect and could be replaced by a continuous shaft.

Case 2: Rotation in opposite direction

The case of the output shaft turning in the opposite direction to the input shaft with no power loss is now analyzed. Again it is seen that the input torque is multiplied by the gear ratio to give the output toque. The transmission reaction torque is now given as minus the sum

Design of Mechanical Power Transmissions

of the input and output torques in Figure 5-4. This can also be written as minus the input torque times the quantity g plus one.

power in = power out

$$T_i \omega_i = T_o \omega_o \Rightarrow T_o = T_i \left(\omega_i / \omega_o \right)$$

but $\left(\omega_i / \omega_o \right) = g$ speed ratio

so $T_o = T_i g$ if $g > 1$ then T_o is multiplied

and $T_r = -(T_o + T)_i = -T_i(g+1)$

Direction of T_r is opposite to input shaft rotation and greater than input torque

Figure 5-4: Rotation in opposite direction

Hence for all values of the gear ratio the reaction torque will be opposite in direction to the input torque and of greater magnitude than the case where the input and output shafts are turning in the same direction.

Example Problem

These analyses are now applied to the following specific problem. A gear transmission is to deliver a torque of 500 ft. lbs. at 2500 rpm to the load being driven. The power is being provided by an engine whose peak torque occurs at 3600 rpm with a power output capability of 300hp. Determine the necessary gear ratio and torque required to restrain the transmission. The direction of rotation of the output shaft is unknown.

Assuming the engine will be run at 3600 rpm the gear ratio can be calculated directly as the input speed divided by the output speed or 3600 over 2500 giving a value of 1.44 n Figure 5-5. With the load speed and torque known the power required by the load can also be determined quantitatively. This calculation results in 238 hp. With no power lost in the transmission this must be the output of the power source.

$$g = \left(\omega_i/\omega_o\right) = \left(N_i/N_o\right) = \frac{3600 \text{ rpm}}{2500 \text{ rpm}} = 1.44$$

$$hp = \frac{2\pi N_o T_o}{33000} = \frac{2(3.14)(2500)(500)}{33000} = 238 \text{ hp}$$

$$T_i = \frac{T_o}{g} = \frac{500}{1.44} = 347 \text{ ft lb}$$

Figure 5-5: Gear Ratio, HP and Input Torque

Since this value is less than 300 hp the engine has sufficient capacity to power the load. To determine the transmission restraining torque the input toque must first be calculated. With a gear ratio of 1.44 the input torque is simply the output toque divided by the gear ratio. This gives a value of 347 ft. lbs.

The transmission restraining torque can now be calculated. If the input and output shafts are turning in the same direction the restraining torque is the input torque times the quantity g minus 1 which gives a value of 152 ft. lbs. as shown in Figure 5-6.

If the output shaft is turning in the opposite direction to the input shaft then the restraining torque is given as minus the input torque times the quantity g plus one resulting in 847 ft. lbs. in the opposite direction to the input shaft rotation.

$$T_r = T_i(g-1) = 347(1.44-1) = 152 \text{ ft lb}$$
$$T_r = -T_i(g+1) = -347(1.44+1) = -847 \text{ ft lb}$$

Figure 5-6: Restraining Torque For Same and Opposite Directions of Rotation

Conclusions to be drawn from this analysis:

- A fixed gear ratio of 1.44 will satisfy the load requirements specified.
- Since the direction of the output shaft is not known the design of the restraint system for the transmission should be able to withstand a torque of 847 ft-lb.
- An engine that generates 300hp at 3600 rpm can probably just as well generate 238 hp at this same speed. This would be particular true for an internal combustion engine running at part throttle.
- It is not known whether the torque output with engine generating 238 hp at 3600 rpm will necessarily be the peak torque for that reduced power output.

Effect of Efficiency

Consider what the effect of power losses in the transmission would be on the solution to this problem. The analysis will be repeated assuming that the transmission is only 85 percent efficient due to internal friction losses. Start with the definition of efficiency as the power out divided by power in, see Figure 5-7. With the gear ratio defined as the output speed divided by the input speed the input torque is given by the required output torque divided by the efficiency and the gear ratio.

$$\eta = \frac{\text{power out}}{\text{power in}} = \frac{T_o \omega_o}{T_i \omega_i}$$

$$\frac{\omega_o}{\omega_i} = \frac{1}{g} \quad \text{from fixed kinematics}$$

$$T_i = \frac{T_o}{\eta g} = \frac{500}{(.85)(1.44)} = 408 \text{ ft lb}$$

Figure 5-7: Efficiency and Input Torque

For the problem just considered this results in a required input torque of 408 ft.-lbs. compared to 347 ft.-lbs. when there were no losses.

This increase in input torque will require an increase in the input power. This can be calculated by either using the new output toque and the original speed or by dividing the output power with no losses by the efficiency of the system. In either instance the result is the same as shown in Figure 5-8.

$$hp_i = \frac{2\pi N_i T_i}{33000} = \frac{2(3.14)(3600)(408)}{33000} = 279 \text{ hp}$$

or $\quad hp_i = \frac{hp_o}{\eta} = \frac{237}{.85} = 279 \text{ hp} \quad$ as a check

Figure 5-8: Horsepower Calculation

The required output from the power source is now 279 hp compared to 238 hp with no losses. Converting this power loss of 41 hp to a heat removal rate results in 1781 BTU per min., see Figure 5-9.

Design of Mechanical Power Transmissions

$$hp_i - hp_o = 279 - 238 = 41\,hp$$

or
$$\dot{Q} = \frac{(41\,hp)\left(33000\,\frac{ft\,lb}{hp\,min}\right)}{\left(778\,\frac{ft\,lb}{BTU}\right)} = 1739\,BTU/min$$

Figure 5-9: Heat Loss Calculation

The power loss will also affect the restraining torque required to hold the transmission stationary since the input torque has increased. If the output shaft turns in the same direction as the input shaft the retaining torque is calculated to be 92 ft.-lbs. This is actually less than when there were no losses.

However, if the input and output shafts are turning in opposite directions the restraining torque is calculated to be 908 ft.-lbs. some 61 ft.-lbs. greater than in the no loss case as shown in Figure 5-10.

$$T_r = T_o - T_i = 500 - 408 = 92\,ft\,lb$$
$$T_r = -(T_o + T_i) = -500 - 408 = -908\,ft\,lb$$

Figure 5-10: Torque Calculation

This now becomes the new criteria for the design of the strength of the restraining system for the transmission if the direction of the output shaft is not known. In these examples it was assumed that the system was operating at steady state conditions. If the power source output and load power required vary with speed the analysis becomes more complex. This is taken up in the case study in Chapter 7 in which both the motor and load characteristics are a function of speed and system start up is analyzed.

Design of Mechanical Power Transmissions

Design of Mechanical Power Transmissions

Chapter 6 – Variable Ratio Transmissions

Chapter 6 covers the analysis and operational characteristics of fluid couplings and torque converters as major variable speed transmissions. The unique physical feature of both of these devices is that the output shaft is not connected mechanically to the input shaft. All power is transmitted through the device by the circulation of a fluid inside sealed casings.

Fluid Coupling Operation

Depicted in Figure 6-1 are schematic cutaways drawings of a fluid coupling. The coupling consists of two elements, a vanned input runner or pump connected to the power source and a vanned output runner or turbine connected to the output shaft. Both elements are contained in a sealed oil filled casing with no mechanical connection.

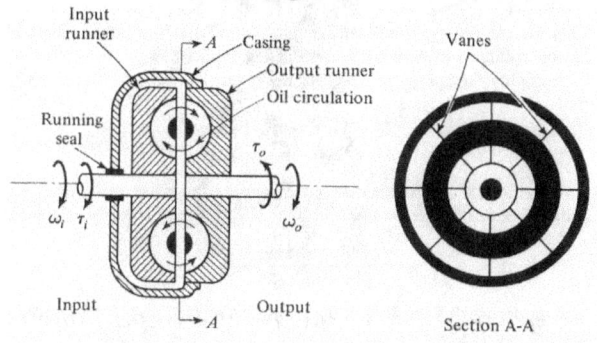

From Weinstein and Angrist

Figure 6-1: 2-D Cutaway of A Fluid Coupling

Rotation of the input runner pump results in centrifugal forces that move the fluid radially outward and across to the turbine vanes of the output element that return it to the pump to be recirculated. This fluid action causes the output turbine and shaft to rotate in the same direction as the input shaft. Since the casing is not physically restrained equilibrium of the system dictates that the output torque is equal to the input torque but they are in opposite directions as indicated. The output shaft speed is less than that of the input shaft. It follows that the output power is less than the input power. The difference between input and output speed divided by the input speed is referred to as slip and is responsible for energy dissipated by internal friction losses.

In this cutaway of an actual fluid coupling in Figure 6-2 the vanes on the input runner pump are clearly visible on the back left side. Although only partially visible the vanes on the output runner are shown attached to the casing housing.

Figure 6-2: 3-D Cutaway Of A Fluid Coupling

The physical separation of the vanes on the input and output runner is visible even though the mechanical separation of the input and output shaft is not as clear.

Design of Mechanical Power Transmissions

Fluid Coupling Analysis

The analysis of the coupling's performance is begun with the definition of slip. This property is simply the difference between the input speed minus the output speed divided by the input speed, see Figure 6-3. It therefore becomes one minus the ratio of the output speed to the input speed. When the output shaft is turning at the same speed as the input shaft the slip is zero and no torque is being transmitted. If the output shaft is held stationary then the slip is one or 100% and the coupling transmits maximum torque. It is assumed that the transmitted or delivered torque is linearly proportional to the slip. Consider the efficiency of the coupling defined as the ratio of the power out to the power in.

$$\text{Slip} = \left(\frac{\omega_i - \omega_o}{\omega_i}\right) = \left(1 - \frac{\omega_o}{\omega_i}\right)$$

Efficiency of coupling

$$\eta = \frac{\text{power output}}{\text{power input}} = \frac{\omega_o T_o}{\omega_i T_i}$$

but from equilibrium $T_o = T_i$

$$\therefore \quad \eta = \frac{\omega_o}{\omega_i} = (1 - \text{slip}) \quad \text{or} \quad \text{slip} = (1 - \eta)$$

Figure 6-3: Slip and Efficiency

Since the input and output torques must be equal the efficiency can be expressed as one minus the slip or conversely the slip is equal to 1 minus the efficiency. When the slip is one or 100% the efficiency of the coupling is zero because the output shaft is not turning. When the slip is zero the efficiency is 100 % but no power is being transmitted because the output torque is zero.

A graph of efficiency as a function of speed ratio or one minus slip is therefore a straight line from the origin to one at slip equal to zero in Figure 6-4. This is actually a plot of efficiency versus efficiency that can only take on this representation.

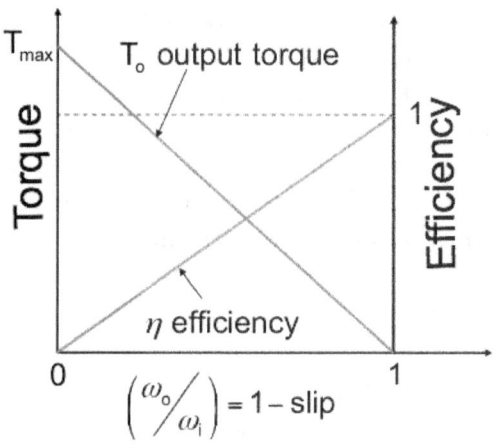

Figure 6-4: Torque and Efficiency Plots

Plotting the torque transmitted on this same graph shows the maximum torque is transmitted at 100% slip or zero efficiency. It decreases linearly to zero torque transmitted when the slip is zero or the efficiency is one because the input and output speeds are equal. This means the power being delivered through the coupling at both zero slip and 100 percent slip is zero.

To get a better understanding of the performance of the fluid coupling consider the power delivered between the two limits previously discussed. The power out is simply the torque out multiplied by the output speed. But the

Design of Mechanical Power Transmissions

torque out can be expressed as T_{max} times the quantity one minus the speed ratio. Now divide both sides of this equation by the maximum power in from the power source expressed as T_{max} times ω_i. The right side then becomes ω_o over ω_i times the quantity one minus ω_o over ω_i. This can also be expressed as the efficiency times the quantity one minus the efficiency, which is a quadratic form. The ratio of power out to max power in will peak at some value on efficiency between 0 and one.

Numerical Example

These operational characteristics will now be used to analyze the application of a fluid coupling in a specific power source and load requirement matching problem. A fluid coupling is proposed as a transmission to smoothly connect intermittently a diesel engine which operates most efficiently at a constant speed of 2500 rpm to a load that requires an input of 150 hp at a speed of 800 rpm. Determine the torque and power being delivered by the diesel engine and the efficiency of the coupling. Calculate how much power is being dissipated as heat.

With the required load speed and specified power source speed given the efficiency of the coupling operation can be calculated directly as 32% as illustrated in Figure 6-5. This low efficiency immediately raises a question as to the appropriateness of using a fluid coupling in this application.

The torque required by the load can now be determined since its power requirement is specified as 150 hp. The coupling output torque must be 985 ft.-lbs. to satisfy the load requirements. Since torque in and torque out for a fluid coupling are equal the power that must be generated by the diesel engine is calculated to be 469 hp.

$$\eta = \frac{800 \text{ rpm}}{2500 \text{ rpm}} = .32 \text{ or } 32\% \text{ (not very high)}$$

Torque required by load

$$\text{from} \Rightarrow T_o = \frac{hp_o(33000)}{2\pi N_o}$$

$$T_o = \frac{(150)(33000)}{2(3.14)(800)} = 985 \text{ ft lb}$$

Power delived by engine note: $T_i = T_o$

$$hp = \frac{2\pi N_i T_i}{33000} = \frac{2(3.14)(2500)(985)}{33000} = 469 \text{ hp}$$

Figure 6-5: Efficiency, Torque and Horsepower Calculation

The power dissipated is simply the difference between the power in minus the power out or the horsepower in multiplied by the quantity one minus the efficiency. Calculating this value either way results in 318 horsepower being dissipated in the coupling, see Figure 6-6. This is more than twice what is being delivered to the load.

The required heat rate removal would be 13,500 BTU per minute, which is exceptionally high. All of these results point to the coupling not being an appropriate device for matching the power source and load requirements as specified. Consider what takes place if the fluid coupling is forced to operate at an efficiency of 80%. The power output from the source would only have to be 188 hp as shown in Figure 6-6.

Design of Mechanical Power Transmissions

$$hp_{dis} = hp_i - hp_o = hp_i(1-\eta)$$

$$hp_{dis} = 468 - 150 = 468(.68) = 318 \text{ hp}$$

or $$\dot{Q} = \frac{(318 \text{ hp})\left(33000 \frac{\text{ft lb}}{\text{hp min}}\right)}{\left(778 \frac{\text{ft lb}}{\text{BTU}}\right)} = 13{,}500 \text{ BTU/min}$$

$$hp_i = \frac{hp_o}{\eta} = \frac{150}{0.80} = 188 \text{ hp}$$

Figure 6-6: Dissipated Power

Assuming that the diesel engine will still be run at 2500 rpm but at part throttle its output torque would drop to 395 ft.-lbs. which is what would be delivered to the load. The output speed of the coupling would be increased to 2000 rpm to provide an 80% efficient operation. These coupling delivery values result in an output torque that is too low and an output speed that is too high for the requirements of the load.

$$T_i = T_o = \frac{hp_i(33000)}{2\pi N_i} = \frac{(188)(33000)}{2(3.14)(2500)} = 395 \text{ ft lb}$$

and the coupling output speed becomes

$$\omega_o = \eta\omega_i = (0.80)(2500) = 2000 \text{ rpm}$$

Figure 6-7: Torque & Speed Outputs

It simply isn't possible to force the coupling to operate at this higher efficiency without some intervention.

Raising the torque delivered to the load and reducing the speed can be accomplished by placing a fixed gear ratio between the coupling and the load. The torque multiplication required would be the ratio of the load requirement to coupling output or 985 ft.-lbs. divided by 395 ft.-lbs. giving a required gear ratio of 2.5, see Figure 6-8. This same result is obtained by dividing the speed requirement of the load into the speed output of the coupling again giving 2.5 to provide proper speed matching.

$$g = \frac{T_{load}}{T_o} = \frac{985}{395} = 2.5 \quad \text{or} \quad g = \frac{N_o}{N_{load}} = \frac{2000}{800} = 2.5$$

Figure 6-8: Gear Ratio Calculation

The operational characteristics of combining a fluid coupling with a fixed gear ratio is accomplished in a torque convertor.

Mechanical Dynamometer Problem

A mechanical dynamometer used to measure power input to some form of test load consists of a fluid coupling and a double compound gear reduction system as shown in Figure 6-9. The gearbox is restrained from rotating about its input and output axes by the addition of an extension at location "p" that rests on a scale that permits the force of constraint to be measured. The numbers of teeth on the double compound gears are listed below the figure.

Design of Mechanical Power Transmissions

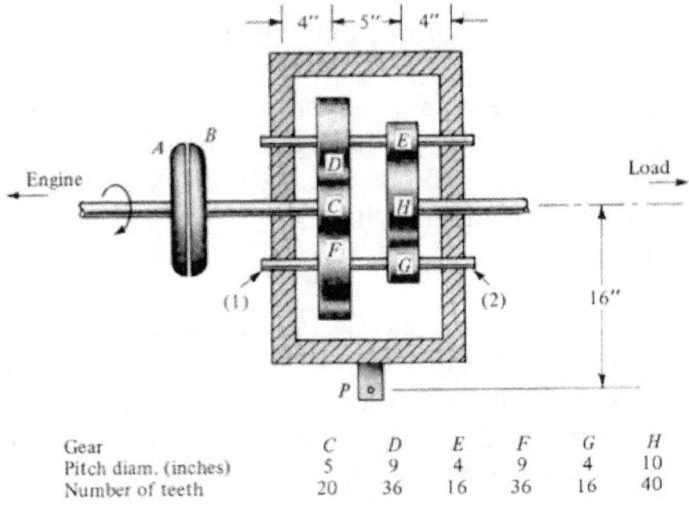

Gear	C	D	E	F	G	H
Pitch diam. (inches)	5	9	4	9	4	10
Number of teeth	20	36	16	36	16	40

Figure 6-9: Mechanical Dynamometer

By measuring the speed of the power source, the load shaft speed and the force exerted at "p" at steady state it is possible to determine the power output of the source and the power delivered to the load. For the specific operating conditions listed in Figure 6-10 these two operational power parameters are to be calculated.

In a specific operation the following conditions apply:

    Engine speed: 2000 rpm
    Load Shaft Speed: 400 rpm
    Force at P: 787.5 lbf (direction unknown)

Figure 6-10: Measured Operating Data

Design of Mechanical Power Transmissions

System Torques and Speeds

The engine speed, ω_E, coupling speed, ω_C, and load speed, ω_L, will all be in the same direction as illustrated in Figure 6-11. This is because of the kinematics of a compound gear train and the lack of constraint for the fluid coupling housing. The engine output torque, T_E, and the coupling output torque, T_C, are in the same direction as the rotation indications since they are both associated with input shafts.

The load torque, T_L, will be in the opposite direction to the output shaft rotation. The direction of the restraining torque, T_T, on the gear housing is assumed to be in the same direction as T_C. If this is incorrect its numerical value will be negative in the calculation to follow.

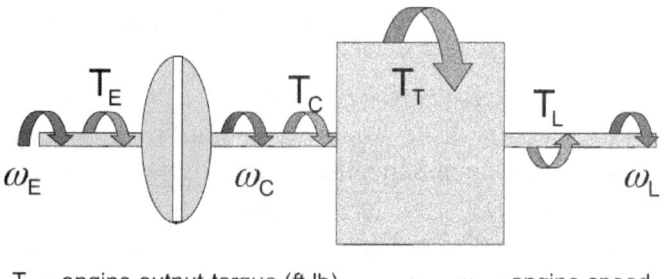

T_E = engine output torque (ft lb)
T_C = coupling output torque (ft lb)
T_L = load torque (ft lb)
T_T = transmission reaction torque (ft lb)

ω_E = engine speed
ω_C = coupling speed
ω_L = load speed

Figure: 6-11 Torques and Speeds

Design of Mechanical Power Transmissions

Transmission Ratio

The gear ratio for a compound gear train is the product of the ratios of the two pairs that form the train. The first pair in this system consists of gears C and D whose numbers of teeth are 20 and 36. Hence gear ratio g_1 is 1.8 as calculated in Figure 6-12.

Compound gear set –

$$g_T = g_1 \times g_2$$

where

$$g_1 = \frac{n_D}{n_C} = \frac{36}{20} = 1.8 \quad \text{and}$$

$$g_2 = \frac{n_H}{n_E} = \frac{40}{16} = 2.5$$

so that

$$g_T = g_1 \times g_2 = 1.8 \times 2.5 = 4.5$$

Figure 6-12: Transmission Ratio

The second pair of gears E and H have teeth numbering 16 and 40 respectively. Their gear ratio, g_2, is 2.5. Therefore the overall gear ratio, g_T is the product of 1.8 times 2.5 or 4.5.

Coupling Efficiency

With the mechanical gear ratio determined and the engine speed and load speed measured, see Figure 6-14, the efficiency of the fluid coupling can be easily determined. The input speed to the gearbox or output speed of the coupling, ω_C, is given by the mechanical gear ratio times the load speed as shown in Figure 6-13. The efficiency, η, of the fluid coupling is then the ratio of the coupling output speed, ω_C, divided by the engine speed, ω_E. In this instance it is 90%.

Input speed to gear box (coupling speed) –

$$\omega_C = g_T \times \omega_L$$
$$\omega_C = 4.5 \times 400 = 1800 \text{ rpm}$$

Coupling Efficiency

$$\eta = \frac{\omega_C}{\omega_E} = \frac{1800}{2000} = 0.9 \text{ or } 90\%$$

Figure 6-13: Coupling Efficiency

Torque Relationships

With the engine speed, ω_E and load speed measured, ω_L, it is necessary to determine the engine torque, T_E, and load torque, T_L, to calculate the engine output power and the load power. This determination is begun with drawing a free body diagram of the gearbox as shown in Figure 6-14.

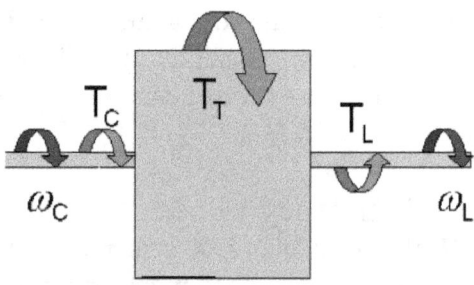

Use to relate torques. Direction of T_T is assumed.

Figure: 6-14 Free Body Diagram of Transmisson

Design of Mechanical Power Transmissions

Applying rotational equilibrium to this diagram results in an equation relating the three torques, T_T, T_C and T_L in Figure 6-15. To solve for these three unknowns torques additional information is required. One added fact is that T_L is related to T_C from the torque multiplication through he gearbox.

Equilibrium applied to transmission unit –
$$\sum T = 0 \Rightarrow T_C + T_T - T_L = 0$$
Torque multiplication through transmission –
$$\frac{T_L}{T_C} = g_T \Rightarrow T_L = g_T T_C$$
Torque across fluid coupling –
$$T_E = T_C$$
Three equations in three unknowns, solve for T_E in terms T_L

Figure 6-15: Torque Relationships

Torque Calculations

The restraining torque, T_T, on the gear case is simply the magnitude of the force measured at point "p" multiplied by its moment arm. In Figure 6-16 this is calculated to be 12,000 in.lbs. Now the first two equations from Figure 6-16 are brought forward and solved simultaneously with T_T of 12,000 in.lbs. to give a value of T_C of 3600 in.lbs. As noted since T_C is a positive umber it means that the restraining torque, T_T, was assumed to act in the correct direction.

Design of Mechanical Power Transmissions

From geometry and data –
$$T_T = F \times R = 787.5 \text{ lb} \times 16 \text{ in} = 12,600 \text{ in lb}$$
and $T_L = g_T T_C = 4.5 T_C$
With
$$T_C + T_T - T_L = 0$$
then
$$T_C - 4.5 T_C = -T_T = -12,600 \text{ in lbs}$$
so that
$$T_C = \frac{12,600}{3.5} = 3600 \text{ in lb}$$
Since T_C is positive direction assumed for T_T was correct

Figure: 6-16 Torque Calculations

Power Calculations

Since the torque across a fluid coupling is the same then T_E is equal to T_C. The engine power can now be calculated since both the speed and torque are known. This is computed in Figure 6-17 as 114 horsepower. Since the efficiency of the fluid coupling was determined earlier it can be used to compute the load power. This simply becomes 90% of the engine power or 103 hp.

Across fluid coupling –
$$T_E = T_C = 3600 \text{ in lb}$$
Power delivered by engine –
$$HP_E = \frac{2\pi N_E T_E}{33000} = \frac{(6.28)(2000)\left(\frac{3600}{12}\right)}{33000} \cong 114 \text{ hp}$$
Efficiency of fluid coupling –
$$\eta = \frac{N_C}{N_E} = \frac{1800}{2000} = 0.9$$
Power delivered to load –
$$HP_L = \eta HP_E = 0.9 \times 114 \cong 103 \text{ hp}$$

Figure 6-17 Power Calculation

Design of Mechanical Power Transmissions

Solution Check

With a torque multiplication through the gearbox of 4.5 the load torque is determined to be 16,200 in. lbs. as shown in Figure 6-19. With that value of load torque and a load speed of 400 rpm the power absorbed by the load is calculated to 103 horsepower which checks the value determined in Figure 6-18.

Start with load torque and speed –

$$T_L = 4.5 T_C = 4.5 \left(\frac{3600}{12} \right) = 1350 \text{ ft lbs} = 16,200 \text{ in lbs}$$

$N_L = 400$ rpm

Power delivered to load –

$$HP_L = \frac{2\pi N_L T_L}{33000} = \frac{(6.28)(400)(1350)}{33000} \cong 103 \text{ hp}$$

Checks with previous calculation

Figure 6-18: Solution Check

Design of Mechanical Power Transmissions

Torque Converter Operation

The physical differences between a fluid coupling and a torque converter are illustrated in this schematic cross section of a torque converter in Figure 6-19. It, too, is a variable speed transmission device in which there is no direct mechanical connection between the input shaft and the output shaft.

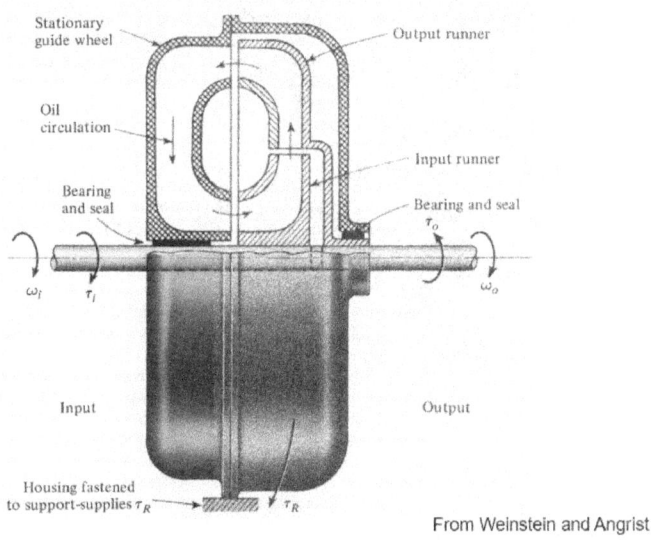

From Weinstein and Angrist

Figure 6-19: Torque Converter Schematic

A torque converter consists of three elements, an input runner or pump, an output runner or turbine and stationary guide vanes. As in the fluid coupling the three elements are not physically connected to one another. Power input to the pump forces fluid thru the vanes of the turbine creating an output torque and shaft rotation in the same direction as the input shaft. The fluid leaving the turbine is directed by the stationary guide vanes back to the pump where its effect is to reduce the power input required to generate a given output torque. An external restraining torque holds the casing and stationary vanes fixed.

From a system view point the torque required to hold the casing and stationary guide vanes fixed is added to the input torque to give the magnitude of the output torque. Thus the device multiplies the torque transmitted. As a consequence of fluid turbulence within the unit energy is dissipated so that the output power is less than the input.

In this cutaway of an actual torque converter all three elements are clearly visible as well as the path of

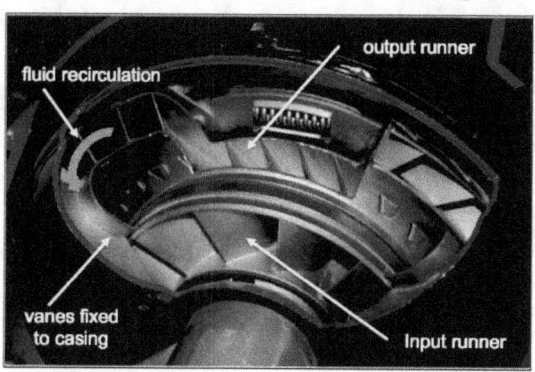

Figure 6-20: Cutaway Section

recirculation of the fluid from the output runner through the fixed vanes back to the input runner. This recirculation returns some of the kinetic energy of the fluid back to the input runner reducing the amount of power required from the power source.

Torque Converter Analysis

There are three performance characteristics that govern the operational behavior of a torque converter. Holding the casing stationary results in the output toque being the sum of the input torque and the casing restraining torque from equilibrium considerations. The efficiency of the device is defined as the power output divided by the power input. In this instance the input and output torques are different from one another and the output speed will be less than the input speed.

$$T_o = T_i + T_r$$

$$\eta = \frac{T_o \omega_o}{T_i \omega_i}$$

$$\frac{T_o}{T_i} = m_t$$

Figure 6-21: Input and Output Torques

Finally the internal vane geometry and orientation of the runners and fixed guides are designed to create a specific torque multiplication.

Numerical Example

These operational characteristics will be applied to the analysis of using a torque converter to match the power source and load requirements of the following problem. A specific load requires an input of 100 hp at 1500 rpm. The torque converter proposed to go between the power source and the load is designed to give a torque output 3.5 times the input torque at zero output shaft speed. At an output shaft speed of 90% of the input speed the delivered torque drops to zero. The output torque can be assumed to vary linearly between these two conditions. The power source can develop a maximum output of 150 hp at 3600 rpm. Determine an appropriate speed and output for the power source and the rate of cooling that must be supplied to the torque converter.

$$\frac{T_o}{T_i} = a + b\left(\frac{\omega_o}{\omega_i}\right)$$

at $\left(\frac{\omega_o}{\omega_i}\right) = 0, \quad \frac{T_o}{T_i} = 3.5 \quad \Rightarrow a = 3.5$

at $\left(\frac{\omega_o}{\omega_i}\right) = .90 \quad \frac{T_o}{T_i} = 0 \quad \Rightarrow b = -\frac{3.5}{.90}$

so that $\frac{T_o}{T_i} = 3.5\left(1 + \frac{1}{.9}\left(\frac{\omega_o}{\omega_i}\right)\right)$ and the effeiciency is

$$\eta = \left(\frac{T_o}{T_i}\right)\left(\frac{\omega_o}{\omega_i}\right) = 3.5\left(\left(\frac{\omega_o}{\omega_i}\right) + \frac{1}{.9}\left(\frac{\omega_o}{\omega_i}\right)^2\right)$$

Figure 6-22: Behavior with Speed

Begin by assuming that the torque output varies linearly with speed. That is T_o is equal to a constant 'a' plus a constant 'b' times the output to input speed ratio as shown in Figure 6-22. One condition used to determine 'a' and 'b' is that when there is no output speed the output torque is 3.5 times the input torque. This is the torque multiplication of the device. A second condition is when the output speed is 90% of the input speed the output torque drops to zero. Applying these two conditions results in 'a' equal to 3.5 and 'b' equal to minus 3.5 divided by .90. The efficiency of the device is obtained by multiplying the torque ratio by the speed ratio. This results in a quadratic equation relating efficiency to the output/ input speed ratio shown in Figure 6-22.

A graphic presentation of the variation of the torque ratio and the system efficiency as a function of the speed ratio is shown in Figure 6-23.

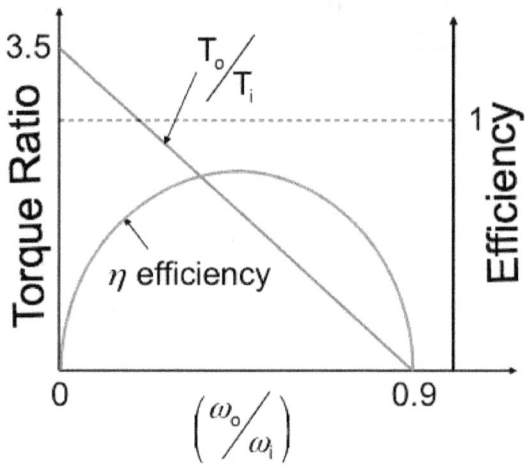

Figure 6-23: Variation of Torque & Efficiency

In this instance the efficiency begins at zero because there is no output speed, goes through a maximum value and then back to zero at ω_o/ω_i equal to .90 since the output torque goes to zero at that point. The behavior of the output to input torque varies linearly from 3.5 to zero as specified by the operating characteristics of the converter.

It will now be assumed that the maximum converter efficiency represents an appropriate operating condition to reduce the power dissipated as heat. To obtain this point of maximum efficiency the derivative of the efficiency function with respect to the speed ratio is set equal to zero. This results in the desired speed ratio being 0.45, which gives an operating efficiency of 78.7% in Figure 6-24.

Design of Mechanical Power Transmissions

This means that only 21.3% of the power delivered by the power source to the converter will be dissipated as heat.

$$\frac{d\eta}{d\left(\omega_o/\omega_i\right)} = 0 = 3.5\left(1 - \frac{2}{.9}\left(\frac{\omega_o}{\omega_i}\right)\right)$$

$$\therefore \quad \left(\omega_o/\omega_i\right)_{opt} = 0.45$$

and $\quad \eta = 3.5\left((4.5) - \frac{1}{.9}(4.5)\right) = 0.787$ or 78.7%

Figure 6-24: Speed Ratio for Maximum Efficiency

By dividing the power requirement of the load by the efficiency, the power out from the power source is determined as 127 hp in Figure 6-25. Also since the selected speed ratio for maximum efficiency was .45 then the output speed of the power source must be set at 3300 rpm. Both of these requirements for the power source are within its capability and close to its maximum output where its efficiency should be high.

$$hp_i = \frac{100}{0.787} = 127 \text{ hp}$$

and $\quad N_i = \frac{1500}{0.45} = 3300 \text{ rpm}$

Figure 6-25: Input Power & Speed

Finally the cooling rate required calculated from the 27 hp dissipated in the converter is only 1145 BTU/min., see Figure 6-26. This is a very reasonable level to be handled possibly with cooling fins on the casing. In conclusion it is seen that the torque converter behaves like the combination of a fluid coupling and a fixed gear train.

Design of Mechanical Power Transmissions

$$\dot{Q} = T_i \omega_i - T_o \omega_o \Rightarrow 127 \text{ hp} - 100 \text{ hp} = 27 \text{ hp}$$

$$\text{or} \quad \dot{Q} = \frac{(27 \text{ hp})\left(33000 \frac{\text{ft lb}}{\text{hp min}}\right)}{778 \frac{\text{BTU}}{\text{ft lb}}} = 1145 \frac{\text{BTU}}{\text{min}}$$

Figure 6-26: Cooling Rate

Design of Mechanical Power Transmissions

Chapter 7 – Matching a Motor to a Load

Introduction

This chapter is a case study covering the details of selecting a simple gear transmission to match an electric motor to a rotating load. Both the power source output and load requirements vary with speed.

Problem Statement

A mixing machine, such as might be used to prepare dough in a bakery, is to be started when full by an electric motor not well suited for the job. The mixer will be driven by a chain drive.

Figure 7-1 Mixing Machine

The torque required by the mixer decreases with speed as indicated in Figure 7-2. The motor is a single-phase ac machine that starts as a repulsion motor and runs as an induction motor. Its starting and running characteristics are given in Figure 7-4. Transfer from "starting" to "running" is affected automatically by a centrifugal switch that may actuate anywhere between 1000 and 1200 rpm. The power input to the mixer at steady running conditions should be as great as possible within limitations of the motor. The problem is to design a suitable gear ratio for the chain drive and determine the power delivered to the mixer at steady operating conditions. How long does it take to bring the mixer to its final speed?

Mixer Characteristics

The operating load characteristics of the mixer are given in terms of the torque required to operate the mixer as a function of speed in rpm. The required starting torque is 36 ft.-lbs. At 1000 rpm only 8 ft. lbs. of torque are required to keep it running at constant speed. This corresponds to a power requirement of 1.5 hp.

Mixing Machine Torque Requirements

Speed (rpm)	Torque (ft. lbs.)
0	36
100	30
200	25
300	21
400	18
500	15
600	12
700	10
800	9
900	8.5
1000	8

Figure 7-2 Mixer Characteristics

Required Mixer Torque

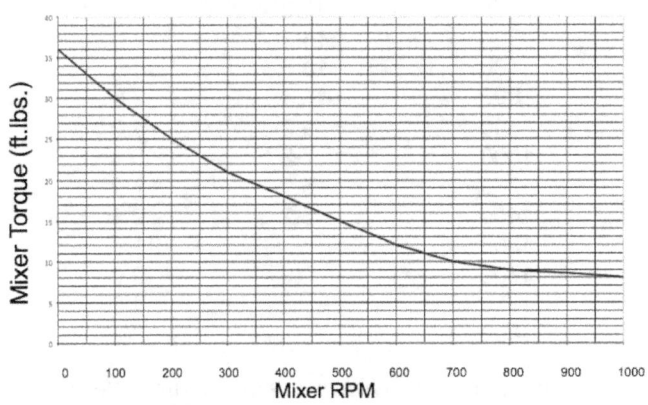

Figure 7-3 Required Mixer Torque

Illustrated in Figure 7-3 is a graph of mixer torque versus speed. It is observed that the torque curve levels off as the mixer approaches it top operating speed. At 50 rpm the required torque is 33 ft. lbs. corresponding to a power requirement of only about 1/3 hp. compared to 1.5 hp. at 1000 rpm. Hence the mixer power requirement is increasing with speed.

Motor Characteristics

Two sets of torque values are given for the motor over its operating speed range. The starting winding in the motor provides high torque initially that reduces significantly as the speed increases. At some speed between 1000 and 1200 rpm a centrifugally actuated switch changes the operation from the starting windings to the running windings.

Delivered Motor Torque

Speed (rpm)	Starting Torque (ft. lbs.)	Running Torque (ft. lbs.)
0	13.0	
200	14.2	
400	13.6	
600	11.3	2.8
800	9.2	4.0
1000	7.2	5.4
1200	6.0	7.2
1400	5.2	9.4
1600		11.2
1700		9.0
1750		4.5
1800		0

Figure 7-4 Motor Torque Characteristics

The running torque begins somewhat lower than the starting winding output but then peaks about 1600 rpm after which it drops off rapidly to its max operating speed of 1800 rpm, see Figure 7-4.

Motor Torque Curves

Illustrated in Figure 7-5 are the two motor torque curves. If switching takes place at 1000 rpm the motor output torque would drop about 30% going from starting to running torque. However, if switching doesn't take place until 1200 rpm the running torque would jump about 20% over the starting torque. It is unknown at what speed the switching from starting to running operation will occur.

Design of Mechanical Power Transmissions

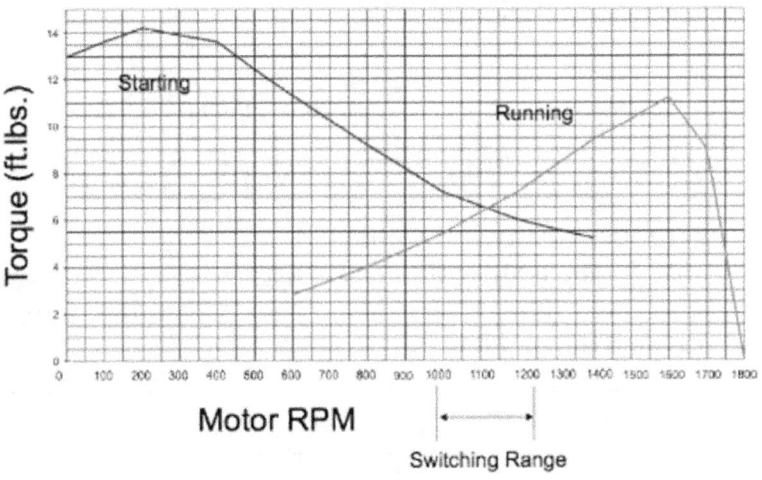

Figure 7-5 Motor Torque Curves

Horse Power Calculation

Determining the selection of an appropriate gear ratio for this problem and calculating the time to come to steady state operating speed will consist of solving several problems. First determine that the power requirements of the mixer can be met by motor output. This is accomplished by calculating power from the equation that the horsepower is equal to 2 π times the rpm N in revolutions per minute times the torque T in ft. lbs. divided by 33,000 ft.-lbs. per minute (Figure 7-6). This is the definition of horsepower in English units. The power delivered by the motor near its top speed should exceed the power required by the mixer at any speed.

Design of Mechanical Power Transmissions

Horsepower definition

$$HP = \frac{2\pi NT}{33,000} \quad \text{where} \quad N = RPM\left(\frac{rev}{min}\right)$$

$$T = torque\,(ft.lbs.)$$

$$HP = 33000\left(\frac{ft.lbs.}{min}\right)$$

Figure 7-6 Horsepower

Numerical Power Values

The table in Figure 7-7 lists the torque and power as a function of speed for the mixer and both the starting and running windings of the motor.

Mixer torque (ft lb)	power (hp)	speed (rpm)	Motor start torque (ft lb)	power (hp)	run torque (ft lb)	power (hp)
36.0	0.00	0	13.00	0.00	0.0	0.00
30.0	0.57	100	13.60	0.26	0.0	0.00
25.0	0.95	200	14.20	0.54	0.0	0.00
21.0	1.20	300	13.90	0.79	0.0	0.00
18.0	1.37	400	13.60	1.04	0.0	0.00
15.0	1.43	500	12.45	1.18	0.0	0.00
12.0	1.37	600	11.30	1.29	2.8	0.32
10.0	1.33	700	10.25	1.37	3.4	0.45
9.0	1.37	800	9.20	1.40	4.0	0.61
8.5	1.46	900	8.20	1.40	4.7	0.80
8.0	1.52	1000	7.20	1.37	5.4	1.03
		1100	6.60	1.38	6.3	1.32
		1200	6.00	1.37	7.2	1.64
		1300	5.60	1.39	8.3	2.05
		1400	5.20	1.39	9.4	2.50
		1500			10.3	2.94
		1600			11.2	3.41
		1700			9.0	2.91
		1800			0.0	0.00

Figure 7-7 Numerical Power Values

Since the motor power output from its running windings near its top speed is greater than any value of power required by the mixer it should be possible to find a gear ratio that will both start the mixer and bring it to some steady state operating speed near its top speed and the top speed of the motor.

Design of Mechanical Power Transmissions

Speed Reduction

It will first prove useful to examine the effects of introducing a gear ratio on the speed of the mixer relative to the motor and the torque required by the mixer as seen by the motor.

Assuming no slippage in the chain drive then from kinematics the product of the radius of the load gear R_L times θ_L, its angular rotation, must be equal to the product of the motor gear radius R_m times θ_m, its angular rotation. Solving for the rotation of the motor gear θ_m gives the ratio of the two radii times the angle of rotation of the load gear. But the ratio of the radii is what is defined as the gear ratio.

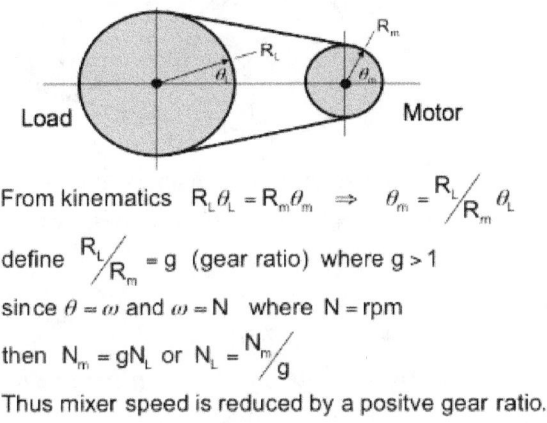

From kinematics $R_L \theta_L = R_m \theta_m \implies \theta_m = \frac{R_L}{R_m} \theta_L$

define $\frac{R_L}{R_m} = g$ (gear ratio) where $g > 1$

since $\theta = \omega$ and $\omega = N$ where $N =$ rpm

then $N_m = g N_L$ or $N_L = \frac{N_m}{g}$

Thus mixer speed is reduced by a positve gear ratio.

Figure 7-8 Speed Reduction

Since θ, angular displacement is proportional to angular speed which is proportional to rpm times 2π then the rpm of the load gear is given by the rpm of the motor gear divided by the gear ratio. Thus mixer speed is reduced by a positive gear ratio.

Design of Mechanical Power Transmissions

Torque Multiplication

From equilibrium considerations the torque T_L being delivered to the load gear is the product of the gear radius R_L times the tension in the chain F. In a similar fashion the torque T_m being delivered by the motor is given the radius R_m times the tension F in the chain. Eliminating the tension F between these two equations gives the ratio of the load torque T_L to the motor torque Tm equal to the gear ratio g. Thus the torque T_L seen by the mixer is equal to the motor torque T_m times the gear ratio g. In other words motor torque is being multiplied by a positive gear ratio as it is delivered to the mixer as illustrated in Figure 7-9.

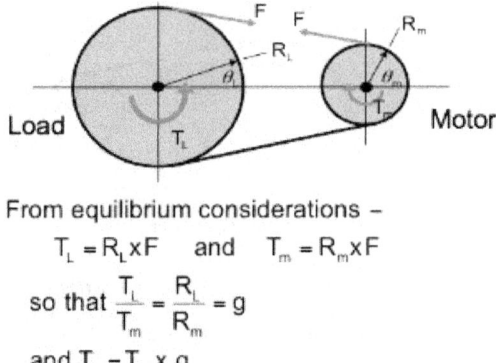

From equilibrium considerations –
$$T_L = R_L \times F \quad \text{and} \quad T_m = R_m \times F$$
so that $\dfrac{T_L}{T_m} = \dfrac{R_L}{R_m} = g$
and $T_L = T_m \times g$

Figure 7-9 Torque Multiplicatio

Torque Required at Motor

Three steps are required to determine if the motor can provide sufficient torque to drive the mixer at any speed for a specific gear ratio. First determine the effective mixer sped N_L for some specified gearing ratio g. This is calculated by dividing the motor speed N_m by the gear ratio. In the second step the motor torque required to match the mixer torque requirement at this reduced mixer speed is determined by dividing the mixer torque T_L at N_L by the gear ratio. This gives the load torque $(T^L{}_m)_m$ that is required from the motor at the motor speed N_m. The third step is to compare the load toque requirement calculated in the second step with the actual value of motor torque T_m at the motor speed N_m. If this required load torque at the mixer is less than T_m then sufficient torque is available to drive the mixer at that speed.

1. Determine the effective mixer speed for the some specified gearing ratio g:

$$N_L = \frac{N_m}{g}$$

2. At N_L the mixer torque is given and designated as T_L. The torque required by the mixer to be delivered by the motor at this speed is then

$$\left(\frac{T_L}{g}\right)_{N_L} = \left(T^L_m\right)_{N_m}$$

3. If $\left(T^L_m\right)_{N_m} \leq T_m$ at N_m then sufficient torque is available to drive mixer.

Figure 7-10: Torque Required

Sample Calculation

Consider the following sample calculation. Assume a gear ratio of three and a motor speed of 600 rpm for which the motor delivery torque Tm is 11.3 ft.-lbs. The mixer speed is 600 rpm divided by 3 or 200 rpm. The torque required at 200 rpm by the mixer is 25 ft. lbs. However, because of the torque multiplication of the gear ratio the mixer only needs to have the motor deliver 25 divided by 3 or 8.33 ft. lbs. as calculated in Figure 7-11.

91

Assume a gear ratio: $g = 3$
Assume a motor speed: $N_m = 600$ rpm
    so that $T_m = 11.3$ ft lbs
then $N_L = N_m/g = 200$ rpm at which $T_L = 25$ ft lbs
but torque delivered to mixer by motor is

$$\left(T_m^L\right)_{N_m} = \left(\frac{T_L}{g}\right)_{N_L} = \frac{25}{3} = 8.33 \text{ ft lbs}$$

since $T_m = 11.3 > 8.33 = \left(T_m^L\right)_{N_m}$

sufficient torque is being supplied to the mixer by the motor. In fact the system will be accereralting.

Figure 7-11 Sample Calculation

Since 8.33 ft.-lbs. is less than 11.3 ft.-lbs. sufficient torque can be delivered by the motor to run the system at a motor speed of 600 rpm with a gear ratio of 3. The difference represents excessive motor torque available to accelerate the system to a higher speed.

Torque Calculations

For motor speeds from 0 to 1800 rpm in increments of 100 the table in Figure 7-12 gives the corresponding calculated mixer speed and required torque to be delivered by the motor for gear ratios of 5, 4.5 and 4 calculated as described in the example in Figure 7-11. As the gear ratio is decreased both the mixer speed and required motor delivery torque are seen to increase.

Motor			Mixer				
	g = 5		g = 4.5		g = 4		
N_m (rpm)	N_L (rpm)	$(T_m^L)_{Nm}$ (ftlb)	N_L (rpm)	$(T_m^L)_{Nm}$ (ftlb)	N_L (rpm)	$(T_m^L)_{Nm}$ (ftlb)
0	0	7.20	0	8.00	0	9.00
100	20	6.96	22	7.70	25	8.63
200	40	6.72	44	7.39	50	8.25
300	60	6.48	67	7.08	75	7.88
400	80	6.24	89	6.78	100	7.50
500	100	6.00	111	6.53	125	7.19
600	120	5.80	133	6.29	150	6.88
700	140	5.60	156	6.04	175	6.56
800	160	5.40	178	5.80	200	6.25
900	180	5.20	200	5.55	225	6.00
1000	200	5.00	222	5.34	250	5.75
1100	220	4.84	244	5.13	275	5.50
1200	240	4.68	267	4.92	300	5.25
1300	260	4.52	289	4.71	325	5.06
1400	280	4.36	311	4.57	350	4.88
1500	300	4.2	333	4.43	375	4.69
1600	320	4.08	356	4.29	400	4.5
1700	340	3.96	378	4.14	425	4.31
1800	360	3.84	400	4.00	450	4.13

Figure 7-12 Torque Calculations

Motor and Mixer Torque

Depicted graphically in figure 7-13 are a series of system torque curves as a function of motor speed. The two upper curves are the starting and running torques of the motor with the centrifugal switch over the range from 1000 to 1200 rpm identified. Also plotted are the mixer torques required of the motor as calculated in Figure 7-12 for gear ratios of 4, 4.5 and 5. For the worst condition of available motor delivery torque, when the motor switches modes at 1000 rpm, a gear ratio of 4.5 will just provide sufficient motor torque to meet required load torque. This is indicated as the critical point. Steady state is reached when the motor torque and load torque are equal at 1750 rpm for the gear ratio of 4.5. For a gear ratio less than 4.5, for example 4.0, the torque required by the mixer will exceed what the motor at 1000 rpm can deliver.

Design of Mechanical Power Transmissions

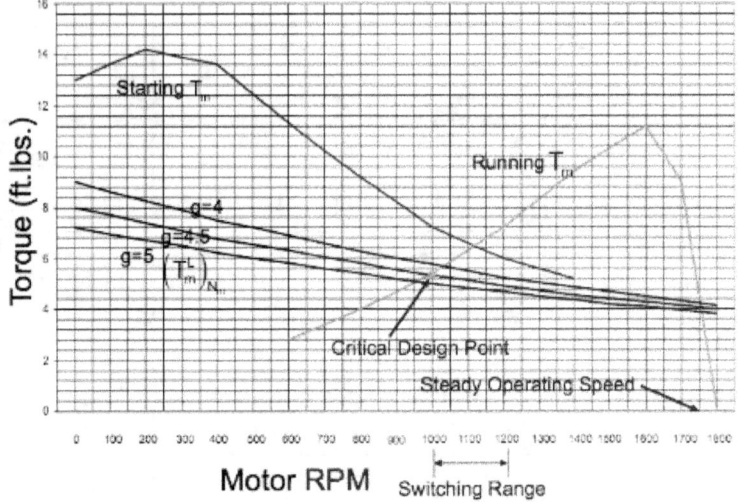

Figure 7-13 Torque Curves

Hence the system would stall and no further acceleration to a final steady state would occur. If the gear ratio is greater than 4.5, for example 5.0, then the final steady speed of operation will be higher than 1750 and the steady state power delivered to the mixer will be less than at the gear ratio of 4.

Evaluation

Assuming no power losses in the chain drive the selection of a gear ratio of 4.5 is the minimum value of g that will insure that the motor drives the mixer to a final operating speed for any switch from starting to running condition for the motor and that the mixer receives maximum power possible. Final operating speed for the motor will be 1750 rpm and 398 rpm for the mixer. The power delivered to the mixer will be given by 2 π times the motor speed of 1750 rpm multiplied by the motor toque of 4.04 ft. lbs. divided by 33,000 ft.-lbs. to give 1.35 hp .

1. Gear ratio of 4.5 is the minimum value of g that will insure that motor drives mixer to final operating speed for any switch from starting to running condition for motor over 1000 rpm and mixer receives maximum power possible.
2. Final operating speed for motor will be 1750 rpm and 398 rpm for mixer.
3. Power delivered to mixer will be

$$HP_{mix} = \frac{2\pi N_m T_m}{33000} = \frac{(6.28)(1750)(4.04)}{33000} = 1.35 \text{ hp}$$

Figure 7-14 Evaluation

System Starting Time

With a gear ratio of 4.5 the motor output torque will always exceed the required mixer torque except at a motor speed of 1000 rpm. This excess torque on the mixer will result in the acceleration of the system until steady state operation is achieved. To determine the time to bring the mixer up to final speed will require a study of the dynamics of the system through an application of Newton's 2nd Law to the motor and the mixer. These equations can then be coupled and solved for the time required for the transient motion during start up.

Newton's second law for rotating bodies states that the net torque acting on the body is equal to its rotational mass moment of inertia times its angular acceleration. This is now applied independently to the motor and the mixer in Figure 7-15. For the mixer this is given by the applied torque from the chain drive F x R_L - T_L equal to the mixer inertia I_L times $d^2\theta_L/dt^2$, where θ_L is the angular displacement of the mixer pulley.

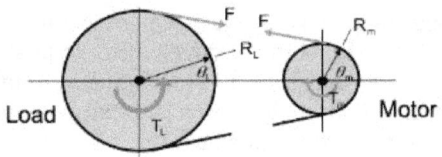

Apply NSL to rotating bodies, i.e. $\sum T = I\alpha$

$$F \times R_L - T_L = I_L \frac{d^2\theta_L}{dt^2}, \quad -F \times R_m + T_m = I_m \frac{d^2\theta_m}{dt^2}$$

Eliminate F between equations

$$T_m - T_L \left(\frac{R_m}{R_L}\right) = I_m \frac{d^2\theta_m}{dt^2} + I_L \left(\frac{R_m}{R_L}\right)\frac{d^2\theta_L}{dt^2}$$

Figure 7-15: Motor/Load Dynamics

For the motor this equation is given by minus the torque from the chain drive $F \times R_m + T_m$ equal to the motor inertia I_m times $d^2\theta_m/dt^2$ where θ_m is the angular displacement of the motor drive pulley. Eliminating the chain tension F between these two equations gives a final single equation of the motion involving θ_m and θ_L together with the motor drive torque T_m and mixer load torque T_L.

Dynamics of the System

It is now recognized that the ratio of R_m to R_L is just equal to one over the gear ratio g. Similarly θ_m and θ_L are also related by the gear ratio g. This simplifies the left side of the dynamics equation to T_m minus T_L over g and the right side to I_m plus I_L over g squared times $d\omega_m/dt$, the angular velocity of the motor. It is further recognized that the left side, $T_m - T_L/g$ is simply the effective accelerating torque of the system T_{eff} effective and the term, $I_m + I_L/g^2$ is the effective inertia of the entire system I_{eff}. In the final

Design of Mechanical Power Transmissions

equation T_{eff} effective is equal to I_{eff} effective times the derivative of the motor speed with respect to time.

$$T_m - T_L \left({R_m}/{R_L} \right) = I_m \frac{d^2\theta_m}{dt^2} + I_L \left({R_m}/{R_L} \right) \frac{d^2\theta_L}{dt^2}$$

but $\left({R_m}/{R_L} \right) = 1/g$ and $\theta_L = \theta_m/g$ so that

$$T_m - {T_L}/{g} = \left(I_m + {I_L}/{g^2} \right) \frac{d^2\theta_m}{dt^2} = \left(I_m + {I_L}/{g^2} \right) \frac{d\omega_m}{dt}$$

now $\left(T_m - {T_L}/{g} \right) = T_{eff}$ (effective accelerating torque)

and $\left(I_m + {I_L}/{g^2} \right) = I_{eff}$ (effective system inertia) so that finally

$$T_{eff} = I_{eff} \frac{d\omega_m}{dt}, \quad \text{units } (ft)(lbs) = (ft\ lbs\ sec^2)\left(\frac{rad}{sec}\right)\left(\frac{1}{sec}\right)$$

Figure 7-16 Dynamics of the System

Integrating for Time

The equation governing the dynamics of the system is now used to determine the time for the mixer to come up to final speed. This is accomplished by first separating variables and then integrating both sides as illustrated in Figure 7-17. On the left the integral of dt from 0 to the final time of t_f is simple t_f. The right side remains I_{eff} times the integral of one over the effective torque T_{eff} integrated with respect to $d\omega_m$ the angular speed of the motor. But ω_m is just equal to 2π times N_m the motor speed in rpm, so the right side integral becomes $2\pi\ I_{eff}$ times the integral of one over T_{eff} integrated with respect to N_m the motor speed in rpm. Finally it is recognized that T_{eff} is simply the value of motor torque minus the mixer torque required geared to that motor speed. Values of this quantity where calculated earlier as a function of motor speed for different gear ratios at increments of one hundred rpm. This permits the value of the integral to be determined numerically.

$$T_{eff} = I_{eff} \frac{d\omega_m}{dt}$$

separate variables and integrate

$$\int_0^{t_f} dt = t_f = I_{eff} \int_0^{\omega_{m'}} \frac{d\omega_m}{T_{eff}}$$

and since $\omega_m = 2\pi N_m$ then

$$t_f = 2\pi I_{eff} \int_0^{N_{m'}} \frac{dN_m}{T_{eff}}, \text{ units sec} = \left(\text{ft lb sec}^2\right)\frac{(\text{rev})}{(\text{sec})}\left(\frac{1}{\text{ft lbs}}\right)$$

recognizing that

$$T_{eff} = \left(T_m - \frac{T_L}{g}\right) = \left(T_m - \left(T_m^L\right)_{N_m}\right)$$

Figure 7-17: Integrating for Time

Set Up for Integration

Before proceeding with the numerical integration for t_f it is necessary to calculate the value of I_{eff}. The inertias of mixer and motor are WR^2 equal to 70 lb. ft^2 for the mixer and WR^2 equal to 5 lb. ft^2 for the motor. The mass moment inertia of the load I_L is determined as its weight W times its radius of gyration R squared divided by g_v the acceleration due to gravity. Numerically this gives 70 divided by 32.2 or 2.177 ft. lb. sec.^{2}. In a similar fashion the inertia for the motor becomes 5 divided by 32.2 or 0.155 ft. lb. sec.^{2}. Finally the effective inertia of the system I_{eff} is the sum of I_m the inertia of the motor plus the inertia of the mixer I_L divided by the gear ratio g squared. This final value of 0.262 ft. lb. sec^2 will be used in performing the integration to get the time required to get to the final speed.

Design of Mechanical Power Transmissions

Numerical Calculations

The table in Figure 7-18 represents the numerical integration of the equation for the time to final speed. The first column is simply motor speed in increments of 100 rpm. The second column is the corresponding mixer speed at the selected gear ratio of 4.5. Two separate limiting cases are considered. Switching that occurs at 1000 rpm and 1200 rpm. At 1000 rpm switching the first column is the available motor torque at each motor speed. The next column is the load torque required by the mixer at that same motor speed. The third column is the difference between these two values, which is the effective torque to produce acceleration. The fourth column is the value of column three times 2π times the effective system inertia multiplied by the change in motor speed for that increment which is 100 rpm. The result is the average time increment required to change the motor speed by 100 rpm.

Motor N_m (rpm)	Mixer M_L (rpm)	T_m (ft lb) Motor switch at 1000rpm	$(T_L)_{lom}$ (ftlb)	T_{eff} (ft lb)	t_{eff} (del N_m)/T_{eff}
0	0	13.00	8.00	5.00	0.50
100	22	13.60	7.70	5.91	0.43
200	44	14.20	7.39	6.81	0.40
300	67	13.90	7.08	6.82	0.40
400	89	13.60	6.78	6.82	0.43
500	111	12.45	6.53	5.92	0.50
600	133	11.30	6.29	5.01	0.59
700	156	10.25	6.04	4.21	0.72
800	178	9.20	5.80	3.40	1.06
900	200	7.30	5.55	1.75	3.03
1000	222	5.40	5.34	0.06	4.46
1100	244	6.30	5.13	1.17	1.59
1200	267	7.20	4.92	2.28	0.93
1300	289	8.30	4.71	3.59	0.65
1400	311	9.40	4.57	4.83	0.51
1500	333	10.30	4.43	5.87	0.43
1600	356	11.20	4.29	6.91	0.47
1700	378	9.00	4.14	4.86	0.52
1750	389	4.50	4.07	0.43	-0.77
1800	400	0.00	4.00	-4.00	
				Sum (sec)	16.87

Figure 7-18: Calculated Starting Times

Design of Mechanical Power Transmissions

The sum of each increment of time in column four up to the final motor speed of 1750 rpm is a good approximation of the total time required for the mixer to come up to its final speed. At 1750 rpm this value is 16.87 seconds. In a similar fashion the last three column calculations are repeated taking into account the switch activates at 1200 rpm. This is not illustrated here but a value of 11.96 seconds is the start up time determined. This is faster than the previous calculated time because the effective torque during the period of motor speed from 1000 rpm to 1200 rpm is greater when the switch activates at 1200 rpm.

Conclusions

The optimum gear ratio to deliver maximum power to mixer is 4.5. The motor will run at 1750 rpm and the mixer at 398 rpm at steady state. Under these conditions the power delivered to the mixer will be 1.35 horsepower. Finally, the time required for the mixer to reach steady state operation will be approximately 12 to 17 seconds depending on when the switch activates.

1. Optimum gear ratio to deliver maximum power to mixer is g = 4.5

2. Motor will run at 1750 rpm and mixer at 398 rpm.

3. Power delivered to mixer at steady state will be 1.35 hp.

4. Time required to come to steady state will be approximately between 12 and 17 seconds.

Figure 7-19: Conclusions

Design of Mechanical Power Transmissions

www.ingramcontent.com/pod-product-compliance
Lightning Source LLC
Chambersburg PA
CBHW060349190526
45169CB00002B/545